Industrial Control and Instrumentation

LIVERPOOL
JOHN MOORES UNIVERSITY
AVRIL ROBARTS LRC
TITHEBARN STREET
LIVERPOOL L2 2ER
TEL. 0151 231 4022

Industrial Control and Instrumentation

W Bolton

 LONGMAN

Addison Wesley Longman Limited
Edinburgh Gate, Harlow,
Essex CM20 2JE, England
and Associated Companies throughout the world.

© Longman Group UK Limited 1991

All rights reserved; no part of this publication
may be reproduced, stored in a retrieval system,
or transmitted in any form or by any means, electronic,
mechanical, photocopying, recording, or otherwise
without either the prior written permission of the Publishers or a
licence permitting resticted copying in the United Kingom issued by
the Copyright Licensing Agency Ltd, 90 Tottenham Court Road,
London, W1P 9HE

First published 1991
Second impression 1995
Third impression 1997

British Library Cataloguing in Publication Data

Bolton, W. (William), *1933—*
  Industrial control and instrumentation
  1. Industries. Process control. Instrumentation
  I. Title
  670.427

Produced through Longman Malaysia, PP

Contents

1 Measurement systems 1
Introduction   1
Measurements system
   elements   1
Performance terms   3
Static and dynamic
   characteristics   8
Laplace transform   11
System transfer
   function   15
System accuracy   16
Sources of error   18
Random and systematic
   errors   18
Noise   19
Intelligent instruments
   19
Calibration   20
Problems   21

2 Control systems   24
Introduction   24
Open- and closed-loop
   systems   25
Basic elements of a
closed-loop system   26
Feedback   29
Closed-loop transfer
   function   30
Open-loop transfer
   function   32
Multi-element, multi-loop
   closed-loop transfer

function   33
Effect of disturbances
   35
Dynamic characteristics
   37
Problems   40

3 Transducers   42
Types of transducers   42
Resistive transducers   45
Capacitive transducers
   52
Inductive transducers   54
Thermoelectric
   transducers   56
Piezo-electric transducers
   59
Photovoltaic transducers
   59
Elastic transducers   59
Pneumatic transducers
   61
Differential pressure
   transducers   62
Turbine transducers   64
Rotating discs   65
Problems   67

**4 Signal conditioning and
   processing   69**
Introduction   69
Wheatstone bridge – null
   method   69

Wheatstone bridge –
deflection type 70
Wheatstone bridge –
thermocouple
compensation 73
Wheatstone bridge –
compensation 74
Alternating-current
bridges 75
Potentiometer
measurement system
78
Signal amplification 79
Signal linearization 84
Voltage to current
converter 85
Current to pressure
conversion 85
Attenuation 86
Filtering 87
Modulation 87
Voltage to frequency
conversion 89
Analogue to digital
conversion 89
Digital to analogue
conversion 91
Sample and hold 92
Multiplexers 93
Problems 93

5 Controllers 95
Introduction 95
Combining data 96
Lag 97
Two-step control 97
Proportional control 100
Derivative control 105
Proportional plus
derivative control 106
Integral control 108
Proportional plus integral
control 110
Proportional plus integral
plus derivative control
112
Cascade control 114
Digital control 116
Tuning 117
Problems 119

6 Correction units 121
The final control
operation 121
Signal conversions 121
Actuators 122
Electric actuators 122
Pneumatic and hydraulic
actuators 124
Control elements 126
Problems 131

7 Data display 132
The range of data-
presentation elements
132
The moving-coil meter
132
The digital meter 134
Alarm indicators 137
Analogue chart recorders
138
Galvanometric recorders
138
Dynamic behaviour of
galvanometric recorders
139
Potentiometric recorders
145
Cathode-ray oscilloscope
146
Monitors 150
Magnetic-tape recorders
151
Digital printers 155
Data loggers 156
Problems 157

8 Measurement systems
159
Designing measurement
systems 159
Temperature-
measurement systems
159
Pressure measurement
167
Flow measurement 172
Measurement of liquid
level 174
Problems 178

9 **Control systems 181**
Designing control systems
    181
Control system
    performance 185
Steady state accuracy 186
Transient response 188
Frequency response 189
Fault finding 194

Problems 194

Appendix
    Laplace transforms 196

Answers to problems 198

Index 201

Preface

The basic aim of this book is to provide an introduction to the principles of industrial control and instrumentation such that:

1   the characteristics and basic principles of elements within such systems can be understood;
2   the ways in which instrumentation and control systems are put together can be appreciated;
3   the reasons for the selection of specific components can be appreciated;
4   manufacturer's specifications can be interpreted and used.

The book is concerned with the basic principles of measurement and control systems and the hardware used. Chapters 1 and 2 give an introduction to the basic concepts and terminology of measurement and control systems. Chapters 3, 4, 5, 6 and 7 offer more detailed discussions of the elements used to build up such systems. Chapters 8 and 9 bring the elements together in a consideration of measurement and control systems.

The aim is seen as being an introduction to the topic and thus the book includes only a superficial consideration of time–variant relationships and their associated mathematics. For the most part the level of mathematics required in this book is just a reasonable acquaintance with algebra. In a few instances an elementary knowledge of calculus is desirable. A basic knowledge of physical science and in particular electrical principles has been assumed.

The book more than covers the Business and Technician Education Council unit 'Industrial Control and Instrumentation U86/345'. This unit has been designed for use within BTEC National Certificate and Diploma courses in the areas of Electrical and Electronic Engineering, Electronic Engineering, Communications Engineering and Computer Engineering though it is also recognised that it might also be appropriate for use in other BTEC programmes. The book is also seen as likely to be of use in other courses, indeed any course where a basic appreciation and introduction to instrumentation and control systems is required.

W. Bolton

1 Measurement systems

Introduction

In engineering, measurement systems are used for essentially three main purposes:

1. to obtain data about some event or item;
2. for inspection or testing, i.e. to determine whether an item is to specification,
3. as an element in a control system.

This chapter is concerned with an overview of the basic elements of all measurement systems, while later chapters consider in more detail the individual elements of such systems.

Measurement systems elements

Fig. 1.1 The input and output signals for the resistance thermometer transducer

Measurement systems can be considered to have three basic constituent elements.

1 *Transducer/sensing element* The *sensing element* or, as it is frequently called, the *transducer* is the first element. This produces a signal which is related to the quantity being measured. For example, a resistance thermometer has a resistor as a transducer (Fig. 1.1). This gives a resistance change when the temperature changes, the greater the change in temperature the greater the resistance change. Sensing elements take information about the thing being measured and change it into some form which enables the rest of the measurement system to give a value to it.

2 *Signal conditioner* The second element is a *signal conditioner* or *signal converter* or *signal processor* which takes the signal from the sensing element and converts it into a condition which is suitable for the display part of a measurement system (Fig. 1.2), or in the case of a control system for combining with the reference signal. An example of this might be an amplifier which takes a small signal from the sensing element and makes it big enough to activate the display. In the

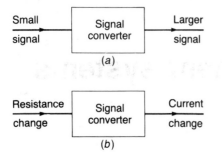

(a)

(b)

Fig. 1.2 Signal conditioners with their input and output signals, (a) an amplifier, (b) an electric circuit for a resistance thermometer

case of the resistance thermometer transducer referred to above, the resistance change produced as a consequence of the change in temperature may be converted to a current change by an electric circuit.

3 *Display* The *display* or *indicating* element is where the output from the measuring system is displayed. This may, for example, be a pointer moving across a scale. In the case of the resistance thermometer with its resistance element in an electric circuit, the output may be displayed on an ammeter. The display element takes the information from the signal converter and presents it in a form which enables an observer to recognise it.

Figure 1.3 shows the general form of a measurement system: a transducer connected to a signal conditioner which in turn is connected to a display element. Figure 1.4 shows a resistance thermometer measurement system, the various parts of which were referred to above.

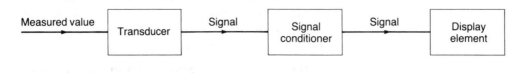

Fig. 1.3 The general form of measurement systems

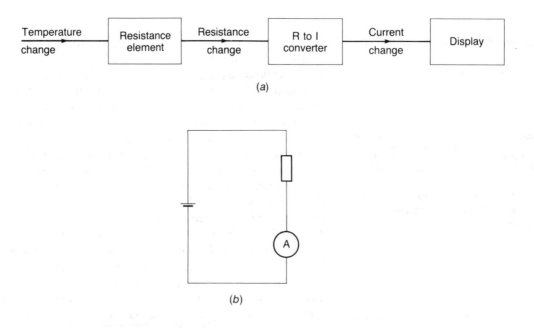

Fig. 1.4 A resistance thermometer (a) block diagram, (b) circuit diagram

Example 1

A thermocouple temperature measuring system uses a thermocouple as the transducer. The output of this transducer is a very small voltage which has to be amplified before it can register on a milliammeter. Draw the block diagram for this system.

Answer

See Fig. 1.5.

Fig. 1.5 A thermocouple temperature measuring system

Performance terms

The following are some of the terms commonly used to describe the performance of measurement systems or elements of such systems.

Accuracy

The *accuracy* of an instrument is the extent to which the reading it gives might be wrong. An ammeter might be quoted as having an accuracy of $\pm 0.2\,A$ when it gives a current reading of $3.0\,A$. This means that the actual current lies somewhere between $2.8\,A$ and $3.2\,A$. Accuracy is often quoted as a *percentage of the full-scale deflection (f.s.d.)* of the instrument. Thus, for example, an ammeter might have a full-scale deflection of $5\,A$ and an accuracy quoted as $\pm 5\%$. This means that the accuracy of a reading of the ammeter between 0 and $5\,A$ is plus or minus 5% of $5\,A$, i.e. plus or minus $0.25\,A$. Hence if the reading was, say, $2.0\,A$ then all that can be said is that the actual current lies between $(2.0 - 0.25)\,A$ and $(2.0 + 0.25)\,A$.

It is often necessary in quoting accuracies to distinguish between whether the measurement is being made of slowly or quickly changing quantities. The term *static accuracy* is used when the quantity being measured is either not changing or changing very slowly, *dynamic accuracy* when it is changing quickly.

Error

The *error* of a measurement is the difference between the result of the measurement and the true value of the quantity being measured.

Error = measured value − true value

Thus if the value given by a thermometer is $25\,°C$ when the true value of the temperature is $24\,°C$ then the error is $+1\,°C$.

If the true value had been 26 °C then the error would have been − 1 °C.

Non-linearity error

For many instruments a linear scale is used. Figure 1.6(*a*) shows an example of a linear scale and Fig. 1.6(*b*) a non-linear scale. A linear scale means the reading given is directly proportional to the distance or angle moved by a pointer across the scale. In many instances however, though a linear scale is used the relationship is not perfectly linear and so errors occur.

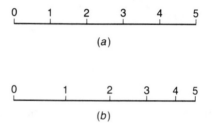

(*a*)

(*b*)

Fig. 1.6 (*a*) A linear scale, (*b*) a non-linear scale

Repeatability

The *repeatability* of an instrument is its ability to display the same reading for repeated applications of the same value of the quantity being measured. Thus, for example, if an ammeter was being used to measure a constant current and gave for four successive readings 3.20 A, 3.15 A, 3.25 A, and 3.20 A then there can be an error with any one reading due to a lack of repeatability.

Precision

Precision is generally used to describe the closeness of the agreement occurring between the results obtained for a quantity when it is measured several times under the same conditions. It is a measure of the scatter of results obtained from measurements as a result of random errors (see later this chapter).

Reliability

The *reliability* of an instrument is the probability that it will operate to an agreed level of performance under the conditions specified for its use. Thus if the specified level of performance of a voltmeter is ± 2% when used at 20 °C then a reliable instrument will give this level of accuracy whenever it is used at 20 °C.

Reproducibility

The *reproducibility* or *stability* of an instrument is its ability to display the same reading when it is used to measure a constant quantity over a period of time or when that quantity is measured on a number of occasions.

Sensitivity

The *sensitivity* of an instrument is given by:

$$\text{sensitivity} = \frac{\text{change in instrument scale reading}}{\text{change in the quantity being measured}}$$

Thus, for example, a voltmeter might have a sensitivity of 1 scale division per 0.05 V. This means that if the voltage being measured changed by 0.05 V then the reading of the instrument will change by one scale division.

Resolution

The *resolution* or *discrimination* of an instrument is the smallest change in the quantity being measured that will produce an observable change in the reading of the instrument. One factor that determines the resolution is how finely the scale of the display is divided into subdivisions. Thus if a thermometer has a scale marked in just 1 °C intervals then it is not possible to estimate a temperature more accurately than about quarter or half a degree.

Range

The *range* of an instrument gives the limits between which readings can be made. For example, an ammeter might have a range of 0–3 A. This means it can be used to measure currents between 0 A and 3 A.

Dead space

The *dead space* of an instrument is the range of values of the quantity being measured for which it gives no reading.

Threshold

When the quantity being measured is gradually increased from zero a certain minimum level might have to be reached before the instrument responds and gives a detectable reading. This is called the *threshold*.

Zero drift

The zero reading of an instrument can change with time. Thus, for example, a meter which might have its pointer on the zero mark on one day might, a month or so later, indicate a small reading despite not having been used to make a measurement.

Lag

When the quantity being measured changes, a certain time, called the *response time*, might have to elapse before the measuring instrument responds to the change. It is said to show *lag*. For example, if a mercury-in-glass thermometer is put into a hot liquid there can be quite an appreciable time lapse before the thermometer indicates the actual temperature.

Hysteresis

Instruments can give different readings for the same value of measured quantity according to whether that value has been reached by a continuously increasing change or a continuously decreasing change. This effect is called *hysteresis* and it occurs as a result of such things as bearing friction and slack motion in gears in instruments. Figure 1.7 shows the type of

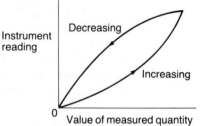

Fig. 1.7 Hysteresis, the readings being different according to whether the measured quantity is increasing or decreasing

relationship that can occur for an instrument showing hysteresis.

Transfer function

The *transfer function* is the ratio of the output of a system, or an element of that system, to its input. The term *gain* is sometimes used instead of transfer function.

$$\text{Transfer function} = \frac{\text{output}}{\text{input}}$$

Thus an amplifier might have a transfer function of 20. This means that the output is twenty times the input. An input of 2 mV would thus give an output of 40 mV. A resistance thermometer element might have a transfer function of $2\,\mu\Omega$ per °C. This means that a change of temperature of 1 °C gives a resistance change of $2\,\mu\Omega$.

Bandwidth

The term *bandwidth* is commonly used to specify the frequency response of amplifiers, though it can be used for other measurement system elements and entire systems. Figure 1.8 shows how the transfer function of a system or element might change with frequency. Generally it is desirable that the response does not vary with frequency. This means that the flat part of the graph is the desirable part. The bandwidth is a measure of the frequency range of this flat part. The bandwidth can be defined as the range of frequencies for which the transfer function is within 70.7% of its peak value G. The 70.7% of G is $G/\sqrt{2}$. An alternative way of expressing

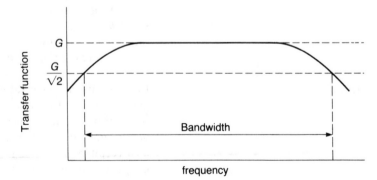

Fig. 1.8 Bandwidth

this is that the bandwidth is the range of frequencies for which the transfer function is within 3 dB (decibels) of its peak value. A change of 3 dB means a transfer function which changes by $1/\sqrt{2}$.

$$\text{Change in dB} = 20 \log_{10}\left(\frac{\text{value}}{\text{max. value}}\right)$$

$$= 20 \log_{10}\left(\frac{1}{\sqrt{2}}\right) = -3$$

Example 2

A voltmeter is specified as having ranges of 0–2 V and 0–20 V, with an accuracy of ± 1% of full-scale deflection. What will be the accuracies of the readings when the voltmeter is used to determine voltages of (a) 1.0 V on the 0–2 V scale and (b) 5.0 V on the 0–20 V scale?

Answer

(a) On the 0–2 V scale the accuracy is ± 1% of 2 V, i.e. 0.02 V. Thus the accuracy of the 1.0 V reading is 1.0 ± 0.02 V.
(b) On the 0–20 V scale the accuracy is ± 1% of 20 V, i.e. 0.2 V. Thus the accuracy of the 5.0 V reading is 5.0 ± 0.2 V.

Example 3

A galvanometer is specified as having a d.c. sensitivity of 4.00 mA per centimetre of scale. What will be the change in reading of such an instrument when the current through it changes by 10 mA?

Answer

The reading will change by 1 cm for each current change of 4.00 mA. Hence a change of 10 mA will mean a reading change of 2.5 cm.

Example 4

A pressure measurement system has a transfer function of 2 scale divisions on the display per pascal change in pressure. What will be the change in display reading when the pressure measured changes by 8 Pa?

Answer

The transfer function for the measurement system is the output divided by the input. Thus for the pressure measurement system

$$2 = \frac{\text{output}}{\text{input}} = \frac{\text{output}}{8}$$

Hence the change in the display, i.e. the output, will be 16 scale divisions.

Example 5

An amplifier is specified as having a bandwidth of d.c. to 25 kHz. What does this mean with regard to the performance of the amplifier?

Answer

The response of the amplifier does not vary more than 3 dB from its maximum transfer function (gain) over the frequency range d.c. to 25 kHz.

Static and dynamic characteristics

The term *static characteristics* of an instrument refer to the steady state reading that it gives when it has settled down. Such characteristics can be expressed in terms of accuracy, linearity, etc. The term *dynamic characteristics* is used to describe the behaviour of an instrument in the time between when the measured quantity changes and a steady reading is given (Fig. 1.9).

To illustrate this consider a thermometer being used to measure the temperature of a liquid. If the temperature of the liquid is suddenly increased then the reading given by the thermometer begins to change. But some time will elapse before the thermometer actually indicates the new temperature of the liquid. Because of this the transfer function for the thermometer varies with time. After a certain time the transfer function reaches a constant value which does not change any further. The transfer function for the thermometer needs to include a term involving time.

Instruments are referred to as *zero-order* instruments when the output or reading is instantaneously reached when there is a change in the measured quantity. The dynamic characteristics of such an instrument are described by an equation of the form

output $\theta_o = k \times$ input θ_i

$\theta_o = k\theta_i$

where k is a constant. The equation does not include any term involving time. An example of such an instrument might be a potentiometer with the output voltage changing immediately the slider is moved along the potentiometer track.

Instruments are referred to as *first order* when the relation between the input and output depends on the rate at which the output changes. A consequence of this is that the output does not instantaneously attain the steady state output but takes some time to reach it. An example of this is a thermometer. For such a system the equation relating the input θ_i and the output θ_o is of the form

$$a_1 \frac{d\theta_o}{dt} + a_o\theta_o = b_o\theta_i$$

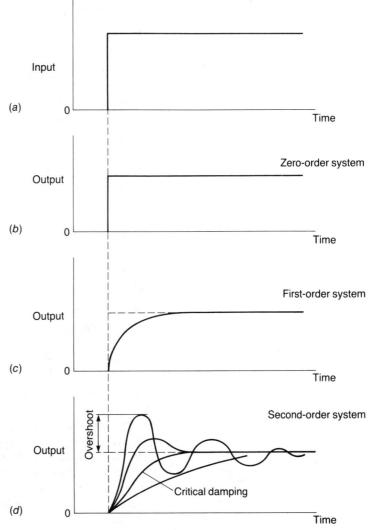

Fig. 1.9 Dynamic responses

where $d\theta_o/dt$ is the rate at which the output changes. a_1, a_o and b_o are constants. When the output has stopped changing the rate term is zero and then $a_o\theta_o = b_o\theta_i$ and there is a simple relationship between input and output, i.e. the steady state transfer function is b_o/a_o.

For a thermometer at temperature T in a liquid at temperature T_1 the rate at which heat enters the thermometer is proportional to the difference between these temperatures, i.e. $(T_1 - T)$. Thus

rate at which heat enters $= k(T_1 - T)$

where k is a constant. This heat will result in a change in temperature of the thermometer. If the thermometer has a specific heat capacity of c and a mass m then a heat input of δQ in a time δt would give a temperature change δT where

$$\delta Q = mc\delta T$$

Hence if dQ/dt is the rate at which heat enters then

$$\frac{dQ}{dt} = mc\frac{dT}{dt}$$

and so

$$k(T_1 - T) = mc\frac{dT}{dt}$$

Rearranged this gives

$$mc\frac{dT}{dt} + kT = kT_1$$

which is in the form stated earlier for the first-order differential equation.

Instruments are referred to as *second order* when the relation between the input and output is of the form

$$a_2\frac{d^2\theta_o}{dt^2} + a_1\frac{d\theta_o}{dt} + a_o\theta_o = b_o\theta_i$$

Such an instrument will take time to reach the steady value and may also oscillate about the steady value for some time. With oscillation the instrument overshoots the steady value. The result depends on the degree of damping in the system. The term *critical damping* is used for the damping which results in an instrument deflecting in the minimum time to the steady value with no overshoot (as shown in Fig. 1.9(*d*). An example of a second-order instrument is a galvanometer (see Chapter 7).

An example of the form of many second-order elements is an elastic sensor which converts a force into a displacement and which consists of a mass m, a spring and damping (Fig. 1.10). The force F applied to the spring is opposed by the force resulting from the stretching or compressing of the spring and that from the damper. The damper can be thought of as a piston moving in a container filled with oil. The force resulting from stretching or compressing the spring is proportional to the change in length of the spring, i.e. the displacement x, hence it can be written as kx with k being the spring stiffness. The force resulting from the damper is proportional to the rate at which the displacement of the piston is varying, i.e. dx/dt, and can be written as cdx/dt where c is a constant. Thus

$$\text{net force applied to mass } m = F - kx - c\frac{dx}{dt}$$

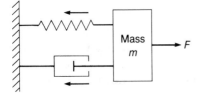

Fig. 1.10 Mass, spring and damper

According to Newton's second law there will be an acceleration due to this net force, with net force equal to mass multiplied by the acceleration. But acceleration is rate of change of velocity and velocity is rate of change of displacement. Hence the acceleration is

$$\text{acceleration} = \frac{dy}{dt} = \frac{d(dx/dt)}{dt} = \frac{d^2x}{dt^2}$$

Hence

$$F - kx - c\frac{dx}{dt} = m\frac{d^2x}{dt^2}$$

This can be rearranged to give

$$m\frac{d^2x}{dt^2} + c\frac{dx}{dt} + kx = F$$

The result is thus a second-order differential equation. A more usual way of writing this equation is as follows. In the absence of the damping a mass m on the end of a spring would freely oscillate with a natural angular frequency ω_n given by

$$\omega_n = \sqrt{(k/m)}$$

A damping ratio ζ is defined as

$$\zeta = \frac{c}{2\sqrt{(mk)}}$$

Thus

$$\frac{k}{\omega_n^2}\frac{d^2x}{dt^2} + \zeta 2\sqrt{(mk)}\frac{dx}{dt} + kx = F$$

$$\frac{1}{\omega_n^2}\frac{d^2x}{dt^2} + \frac{2\zeta}{\omega_n}\frac{dx}{dt} + x = \frac{F}{k}$$

At critical damping the damping ratio ζ has the value 1. For underdamping the ratio is less than 1 and for overdamping greater than 1.

Laplace transform

The transfer function for measurement system elements often needs to include a term involving time. To deal with such forms of transfer function requires more complex mathematics than has been assumed for this text. However it does seem appropriate to give an indication of how systems can be considered when the transfer function of elements in the system depend on time.

As has been indicated above, the relationship between output and input is a differential equation, i.e. an equation involving rates of change. The differential equation is, however, not a very convenient way of seeing how the output depends on the input. One method of simplifying the form of relationship and making it more easy to handle is to use what is called the *Laplace transform* (see Appendix).

For a first-order element the differential equation is of the form

$$a_1 \frac{d\theta_o}{dt} + a_o\theta_o = b_o\theta_i$$

and the corresponding Laplace transform of the equation is

$$a_1 s \times \text{Laplace transform of } \theta_o$$
$$+ a_o \times \text{Laplace transform of } \theta_o$$
$$= b_o \times \text{Laplace transform of } \theta_i$$

$$(a_1 s + a_o) \times \theta_o(s) = b_o \times \theta_i(s)$$

The transfer function $G(s)$ is defined as the ratio of the Laplace transform of the output $\theta_o(s)$ to the Laplace transform of the input $\theta_i(s)$.

$$G(s) = \frac{\text{Laplace transform of output}}{\text{Laplace transform of input}}$$

The transfer function for a first-order element thus becomes:

$$G(s) = \frac{b_o}{a_1 s + a_o}$$

This can be rearranged to give

$$G(s) = \frac{b_o/a_o}{(a_1/a_o)s + 1}$$

b_o/a_o is the steady state transfer function G of the system. a_1/a_o is called the *time constant* τ of the system. Hence

$$G(s) = \frac{G}{\tau s + 1}$$

Consider the behaviour of a first-order element when subject to a step input. The Laplace transform of the output is thus

Laplace transform of output
$$= G(s) \times \text{Laplace transform of input}$$

The Laplace transform for a one unit step input is $1/s$. Hence

$$\text{Laplace transform of output} = G \times \frac{1}{\tau s + 1} \times \frac{1}{s}$$

$$= G \frac{(1/\tau)}{s[s + (1/\tau)]}$$

The transform is of the form

$$\frac{a}{s(s + a)}$$

where $a = (1/\tau)$. We can use tables of Laplace transforms to find out what function would give such a transform (see Appendix). Hence the answer is an exponential growth with time t described by the equation

$$\text{output} = G[1 - \exp(-t/\tau)]$$

Figure 1.11 shows a graph of this equation, showing how the output varies with time for the step input. After a time equal to 1τ the output has risen to just 63% of the final steady output, after 2τ 87%, after 3τ 95%. The final steady value, i.e. 100%, is the steady state transfer function G.

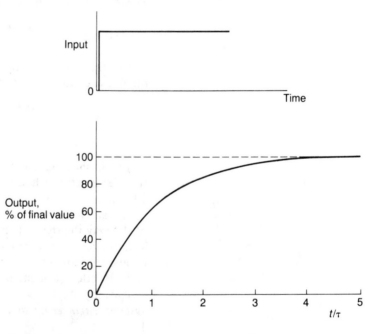

Fig. 1.11 Response of a first-order element to a step input

To illustrate the application of the above, consider the behaviour of a thermocouple at 20 °C when it is first put into a hot liquid at 60 °C. The thermocouple is a first order element and has a transfer function of the form described above. Putting it into the hot liquid is giving the thermocouple a step input of 40 °C. A thermocouple in a sheath has typically a time constant of 2 s if the liquid is not moving. This means that after 2 s the thermocouple output will have risen by 63% of 40 °C, and after 4 s by 87% of 40 °C. Generally about 5τ is needed for a first-order element to reach the steady state condition. This means for this thermocouple about 10 s.

The relationship between the input and output for a second-order element is described by the differential equation

$$a_2 \frac{d^2\theta_o}{dt^2} + a_1 \frac{d\theta_o}{dt} + a_o\theta_o = b_o\theta_i$$

The Laplace transform is

$a_2 s^2 \times$ Laplace transform of θ_o
$\quad + a_1 s \times$ Laplace transform of θ_o
$\quad + a_o \times$ Laplace transform of θ_o
$\quad = b_o \times$ Laplace transform of θ_i

Hence

$$G(s) = \frac{b_o}{a_2 s^2 + a_1 s + a_o}$$

This can be rearranged to give

$$G(s) = \frac{(b_o/a_o)}{(a_2/a_o)s^2 + (a_1/a_o)s + 1}$$

For the mass, spring, damper system described in the previous section (Fig. 1.10), the transfer function is

$$G(s) = \frac{(1/k)}{\dfrac{1}{\omega_n^2} s^2 + \dfrac{2\zeta}{\omega_n} s + 1}$$

A graph showing the behaviour of an element with such a transfer function when subject to a step input is shown in Fig. 1.9(d).

The above discussion is intended to give only a general idea of the way the dynamic behaviour of elements can be treated. The reader is referred to more specialist texts for a more detailed treatment, e.g. *Principles of measurement systems* by J. P. Bentley (Longman, 2nd edn 1988) or *Design of control systems* by A. F. D'Souza (Prentice Hall, 1988) or *Control systems engineering and design* by S. Thompson (Longman, 1989).

Example 6

A bellows (Fig. 1.12) is a transducer which extends when the pressure inside it is increased. The transfer function relating the displacement output and the pressure input is

$$G(s) = \frac{k}{\tau s + 1}$$

where k is a constant for the bellows. If the time constant is $0.5\,\mathrm{s}$, how long will the bellows take to extend fully when the pressure is suddenly increased?

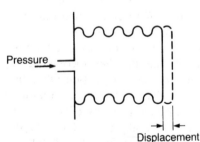

Fig. 1.12 Example 6

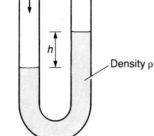

Pressure P

h

Density ρ

Fig. 1.13 Example 7

Answer

For a first-order system the time taken to reach the steady state is about 5τ. Hence for the bellows the time will be about 2.5 s.

Example 7

A U-tube manometer (Fig. 1.13) can be used to measure pressure. The input to the measurement system is the pressure P and the output a difference in height h. The transfer function for such a system is

$$G(s) = \frac{\rho g}{\dfrac{1}{\omega_n^2} s^2 + \dfrac{2\zeta}{\omega_n} s + 1}$$

Describe how the output of the manometer changes with time when there is a sudden change in pressure.

Answer

The manometer is a second-order system. Hence when subject to a step input its output, h, will vary with time according to the graphs in Fig. 1.9(d).

System transfer function

The transfer function G for an entire measurement system is the ratio of the output θ_o to input θ_i.

$$\text{Transfer function } G = \frac{\theta_o}{\theta_i}$$

But a measurement system can be made up of a transducer, signal conditioner and display (Fig. 1.14) and a transfer function can be specified for each of these three elements. The transducer has an input of θ_i and an output to the signal conditioner of θ_1. If the transducer has a transfer function of G_1, then

$$G_1 = \frac{\theta_1}{\theta_i}$$

Fig. 1.14 Transfer function for a measurement system

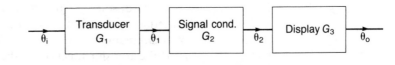

θ_i | Transducer G_1 | θ_1 | Signal cond. G_2 | θ_2 | Display G_3 | θ_o

The signal conditioner has an input of θ_1 and an output of θ_2. If it has a transfer function of G_2, then

$$G_2 = \frac{\theta_2}{\theta_1}$$

The display has an input of θ_2 and an output of θ_o. If it has a transfer function of G_3, then

$$G_3 = \frac{\theta_o}{\theta_2}$$

The equation for the overall transfer function of the measurement system can be written as

$$G = \frac{\theta_o}{\theta_i} = \frac{\theta_1}{\theta_i} \times \frac{\theta_2}{\theta_1} \times \frac{\theta_o}{\theta_2}$$

$$G = G_1 \times G_2 \times G_3$$

The transfer function of the system is equal to the transfer function of the transducer multiplied by the transfer function of the signal conditioner multiplied by the transfer function of the display.

Example 8

What will be the transfer function of a pressure measurement system if the transducer has a transfer function of 0.1 mm per Pa, the signal conditioner a transfer function of 20, and the display a transfer function of 10? (Note: the system is a diaphragm arrangement which deflects when subject to pressure, the signal conditioner then being a lever arrangement which acts as a mechanical amplifier, and the display a pointer moving across a scale.)

Answer

The transfer function of the system is the transfer function of the transducer multiplied by that of the signal conditioner multiplied by that of the display. Hence

System transfer function $= 0.1 \times 20 \times 10 = 20$ mm/Pa

System accuracy

Consider a measurement system which has a transducer, a signal conditioner and a display. Each of these constituent elements will have a certain accuracy and so each will contribute some error to that of the overall system. The question is – what will be the accuracy of the entire measurement system?

If the transfer function of the transducer is G_1, the input θ_i and the output signal θ_1, then in the absence of any error we would have

$$\theta_1 = G_1 \theta_i$$

But because there are some errors in the system the output will fall within a range of values ($\theta_1 \pm \delta\theta_1$). This will mean that the transfer function G_1 can assume a spread of values and should thus be written as ($G_1 \pm \delta G_1$). Hence the actual relationship between the input and output should be

$$\theta_1 \pm \delta\theta_1 = (G_1 \pm \delta G_1)\theta_i$$

This output signal becomes the input signal to the signal conditioner. If this, because of errors, has a transfer function $(G_2 \pm \delta G_2)$ and gives an output signal $(\theta_2 \pm \delta\theta_2)$, then

$$\theta_2 \pm \delta\theta_2 = (G_2 \pm \delta G_2)(\theta_1 \pm \delta\theta_1)$$
$$= (G_2 \pm \delta G_2)(G_1 \pm \delta G_1)\,\theta_i$$

The output signal from the signal conditioner becomes the input to the display. If this, because of errors, has a transfer function $(G_3 \pm \delta G_3)$ and gives an output signal $(\theta_o \pm \delta\theta_o)$, then

$$\theta_o \pm \delta\theta_o = (G_3 \pm \delta G_3)(\theta_2 \pm \delta\theta_2)$$
$$= (G_3 \pm \delta G_3)(G_2 \pm \delta G_2)(G_1 \pm \delta G_1)\,\theta_i$$

θ_o is the output, and $\delta\theta_o$ the error, from the entire system for the input θ_i.

This equation can be simplified. If small terms are ignored, then

$$\theta_o \pm \delta\theta_o$$
$$= (G_3G_2G_1 \pm G_2G_1\delta G_3 \pm G_3G_1\delta G_2 \pm G_3G_2\delta G_1)\,\theta_i$$
$$= G_3G_2G_1\left(1 \pm \frac{\delta G_3}{G_3} \pm \frac{\delta G_2}{G_2} \pm \frac{\delta G_1}{G_1}\right)\theta_i$$

If there were no errors then we would have

$$\theta_o = G_3G_2G_1\theta_i$$

and thus $G_1G_2G_3$ is the overall nominal gain of the system. Hence, dividing both sides of the equation by θ_o gives

$$1 \pm \frac{\delta\theta_o}{\theta_o} = 1 \pm \frac{\delta G_3}{G_3} \pm \frac{\delta G_2}{G_2} \pm \frac{\delta G_1}{G_1}$$

Hence the fractional uncertainty in the output, i.e. $\delta\theta_o/\theta_o$, is the sum of the uncertainties of each transfer function.

$$\frac{\delta\alpha_o}{\theta_o} = \frac{\delta G_3}{G_3} + \frac{\delta G_2}{G_2} + \frac{\delta G_1}{G_1}$$

This is just stating the error in the output is the sum of the errors in each element in the measuring system. This represents the worst possible accuracy.

Example 9

What will be the accuracy of a measurement system for which the transducer has an accuracy of $\pm 2\%$, the signal conditioner an accuracy of $\pm 1\%$, and the display an accuracy of $\pm 0.5\%$?

Answer

The accuracy of the system is the sum of the errors in each element. Thus

$$\text{Accuracy} = \pm(2 + 1 + 0.5) = \pm 3.5\%$$

Sources of error

The following are some of the more common sources of error that can occur with measurement systems.

1 *Construction errors* These are inherent in the manufacture of an instrument and arise from such causes as tolerances on the dimensions of components, on electrical components used, etc.

2 *Approximation errors* A linear relationship between two quantities is often assumed. This may be an approximation or restricted to a narrow range of values. Thus an instrument may have errors due to a component not having a perfectly linear relationship between input and output signals.

3 *Operating errors* These can result from a variety of causes. In reading the position of a pointer on a scale, due to the scale and pointer not being in the same plane, the reading obtained would depend on the angle at which the pointer is viewed against the scale. This common error is known as a parallax error. In addition there are resolution errors due to the uncertainty that exists in reading an instrument's display.

4 *Environmental errors* These are errors which can arise as a result of environmental effects which are not taken account of, e.g. a change in temperature affecting the value of a resistance.

5 *Ageing errors* A consequence of instruments getting older is that some components may deteriorate and their values change, also a build-up of deposits may occur on surfaces which can affect contact resistances and insulation.

6 *Insertion errors* These are errors which result from the insertion of the instrument into the position to measure a quantity affecting its value. For example, inserting an ammeter into a circuit to measure the current will change the value of the current due to the ammeter's own resistance.

Random and systematic errors

Errors can be classified as being either random or systematic errors. *Random errors* are ones which can vary in a random manner between successive readings of the same quantity. They might be due to operating errors, e.g. instruments being read at different angles and so giving variable parallax errors, or environmental errors, e.g. changes in the temperature of the surroundings affecting the calibration of the instrument. Any one reading which is subject to random errors will be inaccurate. However, to some extent random errors can be overcome by taking repeated readings and calculating an average.

Systematic errors are errors which do not vary from one reading to another. They may be the result of construction or approximation errors and so indicated in the accuracy stated for an instrument by the manufacturer. There are also other sources of systematic errors which will not have been allowed for in the specified accuracy, e.g. a bent meter needle and insertion errors.

Noise

The term *noise* is used for the unwanted signals from unrelated electrical circuits and fields that may be picked up by the measurement system and interfere with the signal being measured. This noise may be internally or externally generated with an instrument. In discussing the performance of electronic instruments the *signal-to-noise ratio* is often specified. This is defined as the ratio of the signal level V_s to the internally generated noise level V_n. It is usually expressed in decibels, i.e.

$$\text{signal-to-noise ratio in dB} = 20 \log_{10} (V_s/V_n)$$

Example 10

A measurement system is specified as having a signal-to-noise ratio of 20 dB, what will be the noise with a signal of 1.0 V?

Answer

Using the equation given above

$$\text{signal-to-noise ratio in dB} = 20 \log_{10} (V_s/V_n)$$
$$20 = 20 \log_{10} (1.0/V_n)$$

Hence $V_n = 0.1$ V.

Intelligent instruments

The term *intelligent* when applied to measurement, or control, systems, means that a microprocessor or computer is used for signal processing. The term *dumb* is applied to conventional measurement systems when no such microprocessor is used. Signal processing with a microprocessor requires the signals to be digital rather than analogue. With a digital signal information is transmitted in the form of pulses, i.e. on–off or high–low signals. The off or low signal is represented by the binary number 0 and the on or high signal by the binary number 1. Each such number is called a *bit*. A set of such binary numbers is called a *word*.

Many transducers are analogue devices, i.e. their output is

some replica or scaling of the input. An obvious example of an analogue device is a watch where the time is represented by the position of the watch hands, while the digital watch represents time by a sequence of numbers. Because of this analogue nature of many transducers, intelligent instruments require a signal-conditioning element which converts the analogue transducer ouput to digital form before it can be fed to the microprocessor. If the output from the microprocessor is to be displayed by an analogue system, e.g. a pointer moving across a scale, rather than as just a series of digits then a digital-to-analogue signal-conditioning element is required.

For a detailed discusion of intelligent measurement systems the reader is referred to the book *Intelligent Instrumentation* by G. C. Barney (Prentice Hall 1988).

Calibration

Calibration is the process of putting marks on a display or checking a measuring system against a standard when the transducer is in a defined environment. To do this standards are required.

The basic standards from which all others derive are the *primary standards*. There are primary standards for mass, length, time, current, temperature and luminous intensity. These are defined by international agreement and are maintained by national establishments, e.g. the National Physical Laboratory in Great Britain and the National Bureaux of Standards in the United States.

The primary standard of *mass* is an alloy cylinder (90% platinum–10% iridium) of equal height and diameter, held at the International Bureau of Weights and Measures at Sèvres in France. The mass is defined as one kilogram. Duplicates of this standard are held in other countries.

The primary standard of *length* is the metre and is defined as the length of path travelled by light in a vacuum during a time interval of 1/299 792 458 of a second.

The primary standard of *time* is the second and this is defined as a duration of 9 192 631 770 periods of oscillation of the radiation emitted by the caesium-133 atom under precisely defined conditions of resonance.

The primary standard of *current* is the ampere and this is defined as that constant current which, if maintained in two straight parallel conductors of infinite length, of negligible circular cross-section, and placed one metre apart in a vacuum, would produce between these conductors a force equal to 2×10^{-7} N per metre of length.

The primary standard of *temperature* is the kelvin (K) and this is defined so that the temperature at which liquid water, water vapour and ice are in equilibrium (known as the triple point) is 273.16 K.

The primary standard of *luminous intensity* is the candela and this is defined as the luminous intensity, in a given direction, of a specified source that emits monochromatic radiation of frequency 540×10^{-12} Hz and that has a radiant intensity of 1/683 watt per unit steradian (a unit solid angle).

The term *secondary standards* is used for standards which derive from the primary standards. There is a chain of standards deriving from the original primary standards. Thus the national standards centre will have working national standards for everyday use, rather than directly use the basic primary standards. Calibration laboratories will have reference standards which have been calibrated against the working national standards. Companies will have equipment which has been calibrated by these centres. They in turn may use such calibrated equipment to check the calibration of instrumentation in everyday use in the company.

Example 11

The following data was obtained from a calibration of a 0–100 mA milliammeter. What are the errors associated with each measurement point?

Standard meter (mA)	0	20.2	40.3	60.1	79.9	99.8
Test meter (mA)	0	20.0	40.0	60.0	80.0	100.0

Answer

The errors are the differences between the measured values and the true values. The errors associated with the test meter readings are thus

Test meter (mA)	0	20.0	40.0	60.0	80.0	100.0
Error (mA)	0	−0.2	−0.3	−0.1	+0.1	+0.2

Problems

1 Explain the significance of the following information given in specifications of instruments.

   (a) A moving iron meter: ranges 0–3 A and 0–30 A, a non-linear scale, accuracy ±1%.

   (b) A galvanometer: sensitivity 15 mm/μA.

   (c) An amplifier: gain 40, bandwidth 20–50 kHz.

   (d) A vapour pressure thermometer: ranges −15 to 35 °C and 20 to 100 °C, a maximum error of ±2% of the full-scale reading.

   (e) A piezo-electric pressure transducer: Frequency response flat within ±5% for 5–20 000 Hz.

   (f) A pyrometer: range 500–3000 °C, accuracy ±0.5% of reading, repeatability ±0.4% of full-scale deflection.

   (g) A multimeter: Ranges d.c. voltage 0–100 mV, 3 V, 10 V, 100 V, 300 V; a.c. voltage 0–3 V, 10 V, 100 V, 300 V; d.c. current 0–1 mA, 10 mA, 100 mA, 1 A, 10 A; a.c. current

0–10 mA, 100 mA, 1 A, 10 A; resistance 0–2 kΩ, 200 kΩ, 20 MΩ. Scales: d.c. and a.c. voltages and currents are linear, resistance is non-linear. Accuracy at 20 °C d.c. ±1% of full-scale deflection; a.c. ±2% of full scale deflection, resistance ±3% of the reading at the centre of the scale.

2 For the multimeter referred to in question 1(*g*), what will be the accuracy of the following readings: (a) 50 mA on the d.c. 100 mA scale, (b) 40 V on the 100 V a.c. range?

3 A Venturi flowmeter has a range of 1–10 l/s and an accuracy of ±1% of full-scale deflection. What will be the accuracy when the instrument is used to measure a flow of 4.0 l/s?

4 A temperature-measurement system has as its output a recorder which gives a plot of how the temperature is varying with time. If the system has a transfer function of 10 mm/°C, what will be the deflection on the recorder paper when the temperature changes slowly by 4 °C?

5 What is the transfer function of a piezo-electric pressure transducer which gives an output of 100 mV for a pressure of 120 kPa?

6 A mercury-in-glass thermometer is a first-order measurement system with a time constant of 5 s. Describe how the reading given by the thermometer will change with time when it is suddenly placed in a hot liquid. About how long will it take for a steady reading to be obtained?

7 Explain how the responses of the systems giving the following transfer functions will vary with time when subject to a step input.

(*a*) 12

(*b*) $\dfrac{12}{4s + 1}$

(*c*) $\dfrac{12}{4s^2 + 2s + 1}$

8 What is the transfer function of a pressure-measurement system which uses a pressure transducer with transfer function 10^{-4} Ω/Pa, a signal conditioner with transfer function 20 mV/Ω, and a display with transfer function 20 mm/mV?

9 What is the transfer function of a temperature-measurement system which uses a thermocouple with a transfer function of 40 μV/°C, an amplifier with a transfer function of 800 and a pen recorder display with a transfer function of 300 mm/V?

10 A measurement system employs a transducer with an accuracy of ±1%, a signal conditioner with accuracy ±2% and a display with accuracy ±1%. What is the accuracy of the system?

11 The following results were obtained from the calibration of a thermometer. Plot a graph showing how the errors vary with the temperature indicated by the thermometer.

Test thermometer °C	0.0	10.0	20.0	30.0	40.0	50.0	60.0
Standard thermometer °C	0.0	10.0	20.1	30.1	40.2	40.2	60.3

Test thermometer °C	70.0	80.0	90.0	100.0
Standard thermometer °C	70.3	80.3	90.3	100.4

12 The output from a thermocouple is determined at a range of temperatures which are determined using a thermometer which

has a calibration accuracy much better than that required for the thermocouple. The following is the data obtained. Plot a calibration graph and comment on the errors that would be introduced if the thermocouple was assumed to give a linear response.

Temperature °C	0	20.0	40.0	60.0	80.0	100.0	120.0	140.0
Thermocouple μV	0	0.80	1.61	2.44	3.27	4.10	4.91	5.73

2 Control systems

Introduction

Your body temperature, unless you are ill, remains almost constant regardless of whether you are in a cold or hot environment. To maintain this constancy your body has a temperature-control system. If your temperature begins to increase above the normal you sweat, if it decreases you shiver. Both these are mechanisms which are used to restore the body temperature to its normal value. This control system is maintaining constancy of temperature.

If you go to pick up a pencil from a bench there is a need for you to use a control system to ensure that your hand actually ends up at the pencil. This is done by observing the position of your hand relative to the pencil and making adjustments in its position as it moves towards the pencil. This control system is controlling the positioning and movement of your hand.

One way to control the temperature of a centrally heated house is for a human to stand near the furnace on/off switch with a thermometer and switch the furnace on or off according to the thermometer reading. That is a crude form of control system using a human as a control element. The more usual control system has a thermostat which automatically, without the intervention of a human, switches the furnace on or off. This control system is maintaining constancy of temperature.

Control systems are widespread, not only in nature and the home but also in industry. There are many industrial processes and machines where control, whether by humans or automatically, is required. Control systems can be considered to fall into two main categories. One is *process control* where such things as temperature, liquid level, fluid flow, pressure, etc., are maintained constant. Thus in a chemical process there may be a need to maintain the level of a liquid in a tank to a particular level or to a particular temperature. The other form of control, called a *servo system*, involves consistently and accurately positioning some moving part or maintaining a

24

constant speed. This might be, for example, a motor designed to run at a constant speed or a machining operation in which the position, speed and operation of a tool is automatically controlled.

Open- and closed-loop systems

There are two basic forms of control system, one being called *open-loop* and the other *closed-loop*. The difference between these can be illustrated by a simple example. Consider an electric fire which has a selection switch which allows a 1 kW or a 2 kW heating element to be selected. If a person used the fire to heat a room, he or she might just switch on the 1 kW element if they want the room to be at not too high a temperature. The room will heat up and reach a temperature which is only determined by the fact the 1 kW element was switched on and not the 2 kW element. If there are changes in the conditions, perhaps someone opening a window, there is no way the heat output can be adjusted to compensate. This is an example of open-loop control in that there is no information fed back to the element to adjust it and maintain a constant temperature. The heating system with the electric fire could be made a closed-loop system if the person has a thermometer and switches the 1 kW and 2 kW elements on or off to maintain the temperature of the room constant. In this situation there is feedback, the input to the system being adjusted according to whether its output is the required temperature. This means that the input to the switch depends on the deviation of the actual temperature from the required temperature, the difference between determined by a comparison element – the person in this case. Figure 2.1 illustrates these two types of systems.

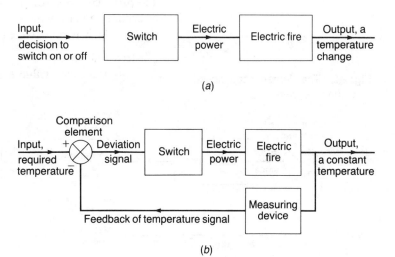

Fig. 2.1   Heating a room, (a) an open-loop system (b) a closed-loop system

To illustrate further the differences between open- and closed-loop systems, consider a motor. With an open-loop system the speed of rotation of the shaft might be determined solely by the initial setting of a knob which affects the voltage applied to the motor. Any changes in the supply voltage, characteristics of the motor as a result of temperature changes or shaft load, will change the shaft speed and not be compensated for. There is no feedback loop. Whereas with a closed-loop system the initial setting of the control knob will be for a particular shaft speed and this will be maintained by feedback, regardless of any changes in supply voltage, motor characteristics or load. In an open-loop control system the output from the system has no effect on the input signal. In a closed-loop control system the output does have an effect on the input signal, modifying it to maintain an output signal at the required value.

Open-loop systems have the advantage of being relatively simple and consequently low cost with generally good reliability. However, they are often inaccurate since there is no correction for error. Closed-loop systems have the advantage of being relatively accurate in matching the actual to the required values. They are, however, more complex and so more costly with a greater chance of breakdown as a consequence of the greater number of components.

Basic elements of a closed-loop system

Figure 2.2 shows the general form of a basic closed-loop system. It consists of the following elements:

1 *Comparison element* The comparison element compares the required or reference value of the variable condition being controlled with the measured value of what is being achieved and produces an error signal. It can be regarded as adding the reference signal, which is positive, to the measured value signal, which is negative in this case.

Error signal
= reference value signal − measured value signal

Fig. 2.2 The elements of a closed-loop control system

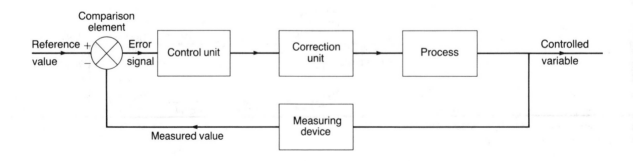

2 *Control element* The control element decides what action to take when it receives an error signal.
3 *Correction element* This sends a signal to the process to produce a change which corrects the controlled condition.
4 *Process element* The *process* is what is being controlled.
5 *Measurement element* The measuring element produces a signal related to the variable condition being controlled.

With the closed-loop system illustrated in Fig. 2.1 for a person controlling the temperature of a room, the various elements are:

Controlled variable   – the room temperature.
Reference value   – the required room temperature.
Comparison element – the person comparing the measured value with the required value of the temperature.
Error signal   – the difference between the measured and required temperature.
Control unit   – the person.
Correction unit   – the switch on the fire.
Process   – the electric fire.
Measuring device   – a thermometer.

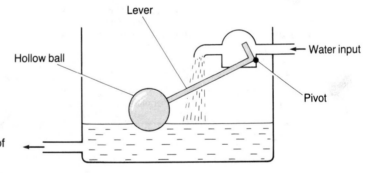

Fig. 2.3 The automatic control of water level in a tank

Figure 2.3 shows an example of a simple control system used to maintain a constant water level in a tank. The reference value is the initial setting of the lever-arm arrangement so that it just cuts off the water supply at the required level. When water is drawn from the tank the float moves downwards with the water level. This causes the lever arrangement to rotate and so allow water to enter the tank. This flow continues until the ball has risen to such a height that it has moved the lever arrangement to cut off the water supply. It is a closed-loop control system with the elements being:

Controlled variable   – water level in tank.
Reference value   – initial setting of lever position.
Comparison element – the lever.

Error signal — the difference between the actual and initial setting of the lever positions.

Control unit — the pivoted lever.

Correction unit — the flap opening or closing the water supply.

Process — water in the tank.

Measuring device — the floating ball and lever.

Figure 2.4 shows a simple automatic control system for the speed of rotation of a shaft. The potentiometer is used to set the reference value, i.e. what voltage is supplied to the differential amplifier as the reference value for the required speed of rotation. The differential amplifier is used both to compare and amplify the difference between the reference and feedback values, i.e. it amplifies the error signal. The amplified error signal is then fed to a motor which in turn adjusts the speed of the rotating shaft. The speed of the rotating shaft is measured using a tacho-generator, connected to the rotating shaft by means of a pair of bevel gears. The signal from the tacho-generator is then fed back to the differential amplifier.

Fig. 2.4 Automatic shaft speed control

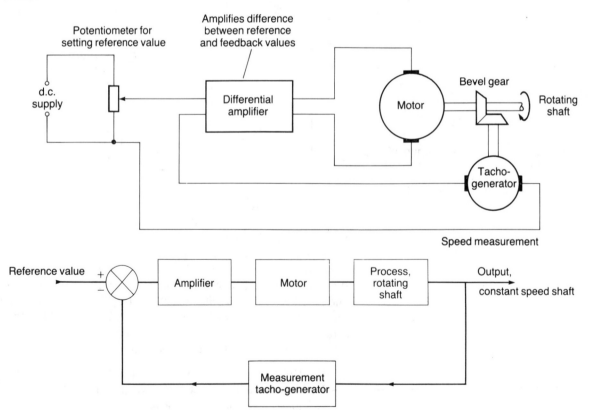

Example 1

A workman maintains the level of the liquid in a container at a constant level. He does this by observing the level of the liquid through a gauge glass in the side of the tank and adjusts the amount of liquid entering it by opening or shutting a control valve. For such a control system what is (*a*) the controlled variable, (*b*) the reference value, (*c*) the comparison element, (*d*) the error signal, (*e*) the control unit (*f*) the correction unit, (*g*) the process, (*h*) the measuring device?

Answer

(*a*) The controlled variable is the level of the liquid in the tank.
(*b*) The reference value is the required level, probably indicated by some mark on the gauge glass.
(*c*) The comparison element is the workman observing the gauge glass.
(*d*) The error signal is the difference between the required level and the actual level indicated through the gauge glass.
(*e*) The workman.
(*f*) The valve.
(*g*) The process is the level of water in a container.
(*h*) The measuring device is the gauge glass with the workman.

Feedback

A *feedback loop* is a means whereby a signal related to the actual condition being achieved is fed back to modify the input signal to the process. The feedback is said to be *negative feedback* when the signal which is fed back is used to reduce the difference between the reference value and the actual value of the controlled variable. With negative feedback

error signal = reference value − feedback signal

Positive feedback occurs when the signal fed back increases the difference between the reference and actual values, i.e.

error signal = reference value + feedback

With control systems the feedback signal is combined with the reference value at the comparison element. This is denoted by the symbol shown in Fig. 2.5 with the reference value being marked as a positive signal and the feedback signal as negative when there is negative feedback and positive when positive feedback.

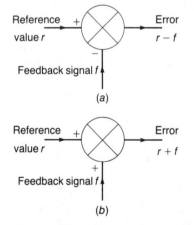

(*a*)

(*b*)

Fig. 2.5 The comparison element with (*a*) negative feedback and (*b*) positive feedback

Example 2

An electrical resistor has the characteristic that the higher its temperature the lower its resistance. When a current passes through the resistor it becomes warm. A consequence of this is that the resistance decreases. This results in the current increasing. The

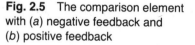

increased current causes the resistor to become even warmer. A consequence of this is that its resistance decreases even further. This results in an increase in current and so on. Is this an example of negative or positive feedback?

Answer

It is positive feedback since the input, the current, is increased by the feedback from the output rather than maintained at a constant value.

Example 3

A closed-loop control system is used to accurately position components in a production operation. The amplifier-valve-positioner part of the system gives 12 mm displacement per millivolt change in input. The feedback loop is a measurement system which gives 0.08 mV per millimetre change in displacement. Figure 2.6 shows the arrangement. What will be the instantaneous error signal produced when the reference signal is suddenly changed by 10 mV?

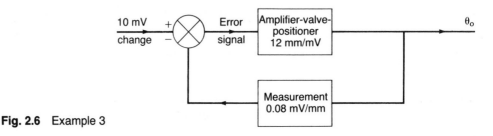

Fig. 2.6 Example 3

Answer

If θ_o is the output then the feedback signal with negative feedback will be $0.08\theta_o$ and so the error signal is

error signal $= 10 - 0.08\theta_o$ mV

This error signal is the input to the amplifier-valve-positioner elements and since this gives an output of 12 mm/mV then the output θ_o must be given by

$$\theta_o = 12(10 - 0.080\theta_o)$$

Hence θ_o is 61 mm and the error signal will be

error signal $= 10 - 0.080 \times 61 = 5.12$ mV

Closed-loop transfer function

The term *transfer function* is defined as being the ratio of the output to input for a system. Figure 2.7 shows a simple closed-loop system. If θ_i is the reference value, i.e. the input, and θ_o the actual value, i.e. the output, of the system then the transfer function of the entire control system is

$$\text{transfer function} = \frac{\text{output}}{\text{input}} = \frac{\theta_o}{\theta_i}$$

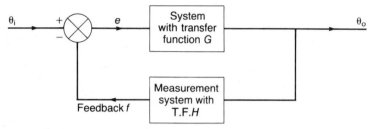

Fig. 2.7 The transfer function with a closed-loop system

Each subsystem within the overall system has its own transfer function. Thus if the system being controlled has a transfer function G then with its input of the error signal e and output of θ_o,

$$G = \frac{\theta_o}{e}$$

If the feedback path has a transfer function H then with its input of θ_o and output f,

$$H = \frac{f}{\theta_o}$$

The error signal e is the difference between θ_i and f, the feedback signal f being a measure of the output of the entire system.

$$e = \theta_i - f$$

Thus substituting for e and f using the above two equations,

$$\frac{\theta_o}{G} = \theta_i - H\theta_o$$

$$\theta_o \left(\frac{1}{G} + H \right) = \theta_i$$

$$\theta_o \left(\frac{1 + GH}{G} \right) = \theta_i$$

Hence the overall transfer function of the closed-loop control system is

$$\text{transfer function} = \frac{\theta_o}{\theta_i} = \frac{G}{1 + GH}$$

The above equation is for negative feedback. With positive feedback the denominator of the equation becomes $(1 - GH)$.

With the closed-loop system, G is termed the *forward-path transfer function* since it is the transfer function relating to the signals moving forward through the system from input to output.

Example 4

A speed-controlled motor has an amplifier-relay-motor system with a combined transfer function of 600 rev/min per volt and a feedback-loop measurement system with a transfer function of 3 mV per rev/min, as illustrated in Fig. 2.8. What is the transfer function of the total system?

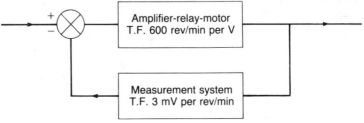

Fig. 2.8 Example 4

Answer

The system will have negative feedback and so the overall transfer function is given by

$$\text{transfer function} = \frac{G}{1 + GH}$$

$$= \frac{600}{1 + 600 \times 0.003}$$

$$= 214 \text{ rev/min per volt}$$

Open-loop transfer function

There are many situations where the transfer function is required for a number of elements in series, with no feedback loop. Consider three components in series, as in Fig. 2.9. It is an open-loop system since there is no feedback loop.

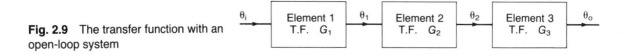

Fig. 2.9 The transfer function with an open-loop system

For element 1 the transfer function G_1 is the output θ_1 divided by the input θ_i. Thus

$$G_1 = \frac{\theta_1}{\theta_i}$$

For element 2 the transfer function G_2 is the output θ_2 divided by its input θ_1. Thus

$$G_2 = \frac{\theta_2}{\theta_1}$$

For element 3 the transfer function G_3 is the output θ_o divided by its input θ_2. Thus

$$G_3 = \frac{\theta_o}{\theta_2}$$

The overall transfer function of the system is the output θ_o divided by the input θ_i. But this can be written as

$$\frac{\theta_o}{\theta_i} = \frac{\theta_1}{\theta_i} \times \frac{\theta_2}{\theta_1} \times \frac{\theta_o}{\theta_2}$$

Hence, for the open-loop system

$$\text{transfer function} = G_1 \times G_2 \times G_3$$

The overall open-loop transfer function is the product of the transfer functions of the individual elements. This applies however many elements there are connected in series.

Example 5

The measurement system used with a control system consists of two elements, a sensor and a signal conditioner (Fig. 2.10). If the sensor has a transfer function of $0.1\,\text{mA/Pa}$ and the signal conditioner a transfer function of 20, what will be the overall transfer function of the measurement system?

Fig. 2.10   Example 5

Answer

The sensor and the signal conditioner are in series so the combined transfer function of the two elements is the product of the transfer functions of the individual elements.

$$\text{Transfer function} = 0.1 \times 20 = 2\,\text{mA/Pa}$$

Multi-element, multi-loop closed-loop transfer function

Consider the closed-loop system shown in Fig. 2.11. The transfer function for the entire system can by obtained by firstly determining the transfer function for the three elements in series. Since these have transfer functions G_1, G_2 and G_3, then their combined transfer function is

Fig. 2.11   The transfer function with a multi-element closed-loop system

$$\text{T.F. of series elements} = G_1 \times G_2 \times G_3$$

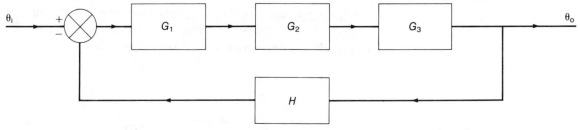

The closed-loop system of Fig. 2.11 can now be replaced by an equivalent simpler system, as in Fig. 2.12. It is now just a single element with a transfer function of $G_1 \times G_2 \times G_3$ and a feedback loop with a transfer function H. The overall transfer function for this system is thus

$$\text{T.F. of system} = \frac{\theta_o}{\theta_i} = \frac{G_1 \times G_2 \times G_3}{1 + (G_1 \times G_2 \times G_3)H}$$

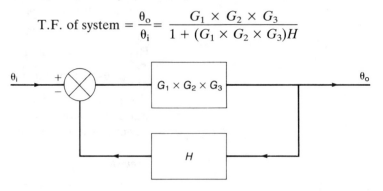

Fig. 2.12 The equivalent system for Fig. 2.11

Some control systems can have more than one feedback loop, as for example in Fig. 2.13. This system has two measurement systems to provide feedback. To obtain the overall transfer function for such a system, the first step is to consider just one of the loops. Thus for the loop with a transfer function H_1, the transfer function for this combined with the system with transfer function G is

$$\text{T.F. for first loop} = \frac{G}{1 + GH_1}$$

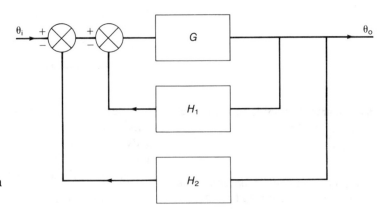

Fig. 2.13 The transfer function with a multi-loop closed-loop system

The system with the two feedback loops can now be replaced by a simpler system with just one loop, as in Fig. 2.14. The overall transfer function for this system is thus

$$\text{T.F. of system} = \frac{G/(1 + GH_1)}{1 + [G/(1 + GH_1)]H_2}$$

$$= \frac{G}{1 + GH_1 + GH_2}$$

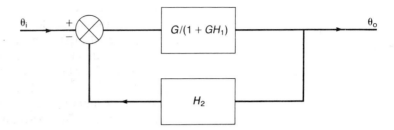

Fig. 2.14 The equivalent system for Fig. 2.13

Example 6

A position-control system used with a machine tool has an amplifier in series with a valve-slider arrangement and a feedback loop with a displacement measurement system (Fig. 2.15). If the transfer functions are as follows, what is the overall transfer function for the control system?

Transfer functions: amplifier 20 mA/V, valve-slider arrangement 12 mm/mA, measurement system 3.0 V/mm.

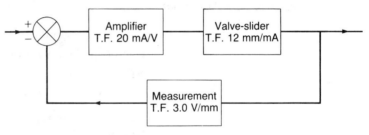

Fig. 2.15 Example 6

Answer

The amplifier and the valve-slider arrangement are in series so the combined transfer function for the two elements is the product of their separate transfer functions.

    T.F. for the series elements = 20 × 12 = 240 mm/V

These elements have a feedback loop with a transfer function of 30 mV/mm. Thus the overall transfer function of the control system is

$$\text{transfer function} = \frac{G}{1 + GH}$$

$$= \frac{240}{1 + 240 \times 0.030}$$

$$= 29 \, \text{mm/V}$$

Effect of disturbances

An important consideration with a control system is the effect of any disturbances. The closed-loop transfer function derived earlier in this chapter indicates how the output relates to the set value. Thus in a domestic heating system if the thermostat setting is changed from 18 °C to 20 °C then the closed-loop transfer function tells us how the output of the control system will change. But suppose we do not change the set value but open a window and let a blast of cold air into the room. How will the control system output react to this change?

We can describe a control system subject to a disturbance like that described above by a block diagram of the form shown in Fig. 2.16. The disturbance gives an input to the process element of the control system, in addition to that from the controller plus correcting unit.

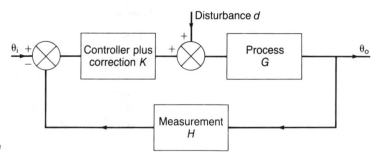

Fig. 2.16 Effect of a disturbance

The signal fed back by the measurement system is $H\theta_o$. This when combined with the set value signal of θ_i means an error signal of

$$\text{error} = \theta_i - H\theta_o$$

After the controller plus correction elements the signal becomes

$$K(\theta_i - H\theta_o)$$

This signal has then the disturbance signal d added to it. Hence the signal entering the process is

$$K(\theta_i - H\theta_o) + d$$

The output θ_o is thus

$$\theta_o = G[K(\theta_i - H\theta_o) + d]$$
$$\theta_o(1 + GKH) = GK\theta_i + Gd$$

and so

$$\theta_o = \frac{GK}{1 + GKH}\theta_i + \frac{G}{1 + GKH}d$$

The first term describes the relationship between the output signal and the set value, the second term that between the output and the disturbance. The effect of a disturbance is thus minimised if the controller gain K is increased.

Dynamic characteristics

The transfer functions of elements used in control systems tend to include a term involving time. This is because they do not respond instantaneously to changes and take time to reach a steady state. Thus, for example, for a system used to control the level of water in a tank, if there is an abrupt drop of level the control system will respond but some time will elapse before the water level returns to its set level. Thus the overall transfer function for the control system must include a term involving time. This is because the transfer functions of the elements in the system include terms involving time.

The following is a discussion of a simple system and is intended to indicate the way in which dynamic control problems can be approached. For further discussion the reader is referred to more specialist texts, e.g. *Control systems engineering and design* by S. Thompson (Longman 1989) or *Design of control systems* by A. F. D'Souza (Prentice Hall 1988).

Consider the control system shown in Fig. 2.17 for the level of water in a tank. The level is controlled by control of the rate at which water enters the tank. If the water enters at the rate of Q_{in} per second and leaves at the rate of Q_{out} per second then the net rate at which the water increases in the

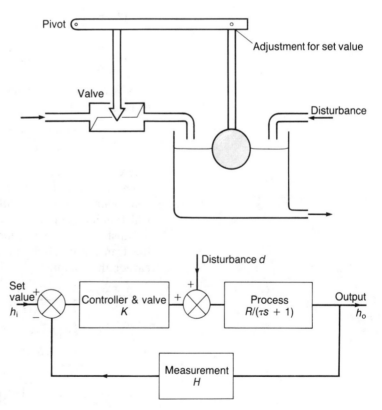

Fig. 2.17 Control of water level

tank is $(Q_{in} - Q_{out})$. In a time δt the change in volume of water in the tank will be $(Q_{in} - Q_{out})\delta t$. If the tank has a cross-sectional area of A then this change in volume will produce a change in water level of δh, where

$$(Q_{in} - Q_{out})\delta t = A\delta h$$

$$Q_{in} - Q_{out} = A\frac{\delta h}{\delta t}$$

Hence

$$Q_{in} - Q_{out} = A\frac{dh}{dt}$$

The outflow from the tank will be affected by the height of the water above the exit pipe, i.e. the pressure head. Since the pressure due to a height of water h is proportional to h we can write

$$Q_{out} = \frac{h}{R}$$

where R represents the hydraulic resistance of the exit pipe. Thus

$$Q_{in} - \frac{h}{R} = A\frac{dh}{dt}$$

$$RA\frac{dh}{dt} + h = RQ_{in}$$

This is a first-order equation and thus has a transfer function of (see Ch. 1):

$$G(s) = \frac{\text{output transform}}{\text{input transform}} = \frac{R}{\tau s + 1}$$

where $\tau = RA$ and is the time constant.

The above is the transfer function for the process. With a controller plus valve with a combined transfer function K which is independent of time and a measurement system with a transfer function H which also is independent of time, then the transfer function for the control system is given by (see earlier this chapter):

$$h_o(s) = \frac{GK}{1 + GKH}h_i(s) + \frac{G}{1 + GKH}d(s)$$

$$= \frac{[R/(\tau s + 1)]K}{1 + [R/(\tau s + 1)]KH}h_i(s)$$

$$+ \frac{[R/\tau s + 1)]}{1 + [R/(\tau s + 1)]KH}d(s)$$

$$= \frac{RK}{\tau s + 1 + RKH}h_i(s) + \frac{R}{\tau s + 1 + RKH}d(s)$$

$h_o(s)$ is the Laplace transform of the output level, $h_i(s)$ is the transform of the set level and $d(s)$ the transform of the disturbance flow.

Example 7

What is the transfer function of the system shown in Fig. 2.18?

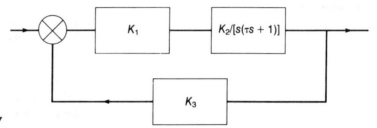

Fig. 2.18 Example 7

Answer

The transfer function of the system can be obtained using the expression derived earlier in this chapter of

$$\text{transfer function} = \frac{G}{1 + GH}$$

The forward transfer function G is $K_1 \times K_2/[s(\tau s + 1)]$, hence

$$\text{transfer function} = \frac{K_1 K_2/[s(\tau s + 1)]}{1 + K_1 K_2 K_3[s(\tau s + 1)]}$$

$$= \frac{K_1 K_2}{s(\tau s + 1) + K_1 K_2 K_3}$$

Example 8

A system with a transfer function $K/(\tau s + 1)$ is subject to a ramp input. State an equation describing how the output of the system will vary with time. Laplace transforms are given in the Appendix.

Answer

For the system

$$G(s) = \frac{\text{output transform}}{\text{input transform}} = \frac{K}{\tau s + 1}$$

Hence, since the Laplace transform of a ramp input is $1/s^2$,

$$\text{output transform} = \frac{K}{\tau s + 1} \times \frac{1}{s^2}$$

$$= \frac{K/\tau}{s^2[s + (1/\tau)]}$$

The Appendix indicates that the equation which would give a Laplace transform

$$\frac{a}{s^2(s + a)}$$

is $t - [(1 - e^{-at})/a]$. Hence

$$
\begin{aligned}
\text{output} &= Kt - K[(1 - e^{-t/\tau})/1/\tau)] \\
&= K(t - \tau) + K\tau e^{-t/\tau}
\end{aligned}
$$

Problems

1 Explain the difference between open- and closed-loop control.
2 Identify the basic elements of a closed-loop control system and represent them on a block diagram of the system.
3 Identify the basic elements in the closed-loop control systems involved in (*a*) a driver steering a car, (*b*) a thermostat-controlled central-heating system.
4 Figure 2.19 shows a system used to control the rate of flow of liquid along a pipe. Explain how the system operates.

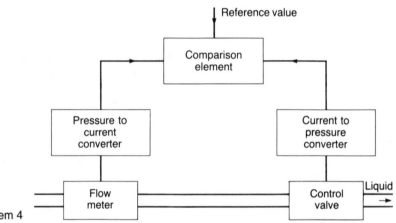

Fig. 2.19 Problem 4

5 Explain the difference between positive and negative feedback and give an example of each.
6 The automatic control system for the temperature of a bath of liquid consists of a reference voltage fed into a differential amplifier. This is connected to a relay which then switches on or off the electrical power to a heater in the liquid. Negative feedback is provided by a measurement system which feeds a voltage into the differential amplifier. Sketch a block diagram of the system and explain how the error signal is produced.
7 A temperature measurement system has a thermometer which produces a resistance change of $0.007\,\Omega/°C$ connected to a Wheatstone bridge which produces a current change of $20\,mA/\Omega$. What is the overall transfer function of the system?
8 For the control system described in problem 4 and illustrated in Fig. 2.19, what will be (*a*) the transfer function for the feedback loop if the flow meter has a transfer function of $2\,kPa$ per m/s and

the pressure to current converter 1.0 mA per kPa,⁻ (b) the transfer function for the forward path if the current to pressure converter has a transfer function of 6 kPa per mA and the control valve 0.1 m/s per kPa, and (c) the overall transfer function of the control system?

9 What are the transfer functions for the control systems in Fig. 2.20 if the elements have the transfer functions specified?

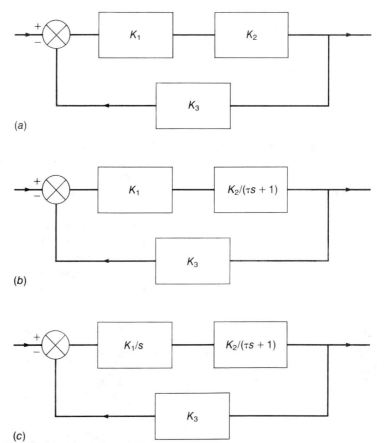

(a)

(b)

(c)

Fig. 2.20 Problem 9

3 Transducers

Types of transducers

The transducer, or sensing element, is the first element in a measuring system and takes information about the variable being measured and transforms it into a more suitable form for the measurement system. The physical principles involved in this transformation are very varied. For example, with a resistance thermometer the resistance depends on its temperature, hence information about the temperature of a hot body is transformed into information in the form of resistance (Table 3.1). Such a transducer is said to be a *passive device* since it does not itself generate any electrical power.

Table 3.1 The resistance thermometer transducer

Input	Physical principle	Output
Temperature change	Resistance depends on the temperature	Resistance change

An *active device* is one where the transducer is itself a source of electrical power. An example of this is a thermocouple. As a consequence of the thermoelectric effect this generates an e.m.f. which is related to the temperature being measured (Table 3.2). Through an appropriate circuit this e.m.f. can be used to give a current through a meter. All transducers which give electrical signal outputs can be classified as either passive or active.

Table 3.2 The thermocouple transducer

Input	Physical principle	Output
Temperature	Thermoelectric effect	e.m.f.

A simple example of the use of a mechanical transducer is a spring balance. This has as its input a force. The transducer is the spring and the physical principle involved is elasticity, the action of forces causes the spring to stretch. The output of the transducer is thus a change in length (Table 3.3).

Table 3.3 The spring balance transducer

Input	Physical principle	Output
Force	Elastic extension of a spring	Change in length

Many measuring systems use more than one transducer. Thus a load cell used for the measurement of force may consist of a cylinder which is elastically deformed by the action of the force with the deformation then being detected by strain gauges which give a resistance change when subject to strain.

In order to give some idea of the range of transducer devices and physical principles involved, Table 3.4 outlines some of the more common forms of transducer. They are grouped according to the form of their output as electrical passive, electrical active and mechanical, with subdivision within the groups according to the types of output concerned. More specific information follows later in this chapter.

Table 3.4 Transducers

Output	Physical principle	Typical applications
Electrical passive		
Resistance	Resistance of an element depends on temperature	Measurement of temperature, flow rate
	Resistance of a wire or semiconductor element depends on the strain	Strain gauge for measurement of strain or in load cells for measurement of force
	The position of the slider of a potentiometer determines that part of its resistance in a circuit	Measurement of displacement
	The resistance of photoconductive materials depends on the intensity of light	Detection and measurement of light intensity

LIVERPOOL JOHN MOORES UNIVERSITY
LEARNING SERVICES

Table 3.4 *contd.*

Output	Physical principle	Typical applications
Capacitance	The capacitance of a parallel plate capacitor depends on the distance between the plates	Measurement of displacement or pressure
	The capacitance of a two-concentric-cylinder capacitor depends on the dielectric between the cylinders	Measurement of liquid level
Inductance	The reluctance of a magnetic circuit is affected by changes in the magnetic flux path	Measurement of displacement or position or force or pressure
	The inductance of a coil depends on the permeability of its core, hence movement of a high permeability rod into a coil changes inductance	Measurement of displacement or position or force
	The difference in the voltages of two secondary coils of a transformer as a result of the position of the iron core	Linear variable differential transformer for measurement of displacement, position, pressure, force
Electrical active		
Thermoelectric	An e.m.f. is produced when there is a difference in temperature between the junctions of two dissimilar metals	Thermocouple
Piezo-electric	An e.m.f. is generated when a force is applied to certain crystals	Measurement of pressure changes, acceleration, vibration, sound
Photovoltaic	A voltage is produced in a semiconductor junction element when radiation is incident on it	Light meter

Table 3.4 *contd.*

Output	Physical principle	Typical applications
Mechanical		
Elasticity	An elastic element, such as a spring or diaphragm or bellows or coil changes in length when subject to a force	Measurement of force, pressure, acceleration, torque
Pneumatic	The pressure of compressed air which is escaping through a nozzle depends on the proximity of a surface to that nozzle	Measurement of displacement
Differential pressure	The difference in pressure between fluid at rest and in motion depends on the fluid velocity	Measurement of fluid velocity
Turbine	The rate at which a turbine in a fluid flow rotates depends on the flow rate of the fluid	Measurement of rate of flow of a fluid
Rotating disc	An e.m.f. is induced when the flux linked by a coil changes as a result of a toothed ferromagnetic wheel rotating near the coil	Measurement of angular velocity
	Optical, electrical and electromagnetic transducers can be used to detect the pulses produced by a rotating disc	Measurement of shaft angular velocity

Resistive transducers

Metal wire resistance thermometers

The resistance of most metals increases at a reasonably linear rate with temperature. For such a linear relationship

$$R_t = R_o(1 + \alpha t)$$

where R_t is the resistance of a length of wire at temperature $t\,°C$ and R_o its resistance at $0°C$. α is called the temperature coefficient of resistance, unit $°C^{-1}$, and depends on the metal concerned. Figure 3.1 shows the resistance–temperature graphs for three commonly used metals.

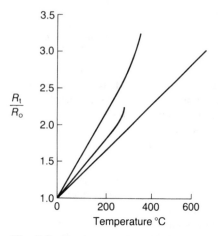

Fig. 3.1 Resistance–temperature graphs for platinum, nickel and copper

Platinum is widely used for *resistance thermometers*. This metal has a closely linear resistance–temperature relationship, gives good repeatability, can be used over a wide range of temperatures (about $-200\,°C$ to $+850\,°C$) and because it is relatively inert can be used in a wide range of environments without deterioration. It is however more expensive than many other metals but the benefits outlined above tend to outweigh the cost factor. The temperature coefficient of resistance α is about $3.9 \times 10^{-3}\,°C^{-1}$

Nickel and copper are cheaper alternatives but are more prone to interaction with the environment and cannot be used over such a large range of temperature. Nickel has a temperature coefficient of resistance α of about $6.7 \times 10^{-3}\,°C^{-1}$ and a range of about -80 to $+300\,°C$. Copper has a temperature coefficient of resistance α of $3.8 \times 10^{-3}\,°C^{-1}$ and a range of about -200 to $+250\,°C$.

A more accurate relationship between resistance and temperature for metals which takes account of nonlinearity is

$$R_t = R_o(1 + \alpha t + \beta t^2 + \gamma t^3)$$

where α, β and γ are temperature coefficients of resistance, with $\alpha > \beta > \gamma$. For platinum α is about $3.9 \times 10^{-3}\,°C^{-1}$, β about $-5.9 \times 10^{-7}\,°C^{-2}$ and γ is generally so small as to be neglected.

Whatever the metal used, the resistance element generally consists of the resistance wire wound over a ceramic coated tube, it then being also coated with ceramic, and mounted in a protecting tube. The result is a probe for immersion in the medium whose temperature is being measured. The response time is fairly slow, often of the order of a few seconds, because of the poor thermal contact between the coil and the medium whose temperature is being measured. The presence of the protecting tube inevitably increases the response time.

Example 1

What is the change in resistance of a platinum resistance coil of resistance $100\,\Omega$ at $0\,°C$ when the temperature is raised to $30\,°C$? The temperature coefficient of resistance can be taken as $0.0039\,°C^{-1}$.

Answer

Assuming that the resistance varies linearly with temperature,

$$R_t = R_o(1 + \alpha t)$$
Change in resistance $= R_t R_o = R_o \alpha t$
$$= 100 \times 0.0039 \times 30 = 11.7\,\Omega$$

Thermistors

As well as metals, *thermistors* are used for temperature measurement. Thermistors are small pieces of material made from mixtures of metal oxides, such as those of chromium, cobalt, iron, manganese and nickel. The material is formed into various forms of element, such as beads, discs and rods (Fig. 3.2). The resistance of thermistors generally decreases with an increase in temperature and is highly non-linear though there are some for which the resistance increases with an increase in temperature. Figure 3.3 shows a typical graph. The change in resistance per degree change in temperature is considerably larger than that which occurs with metals.

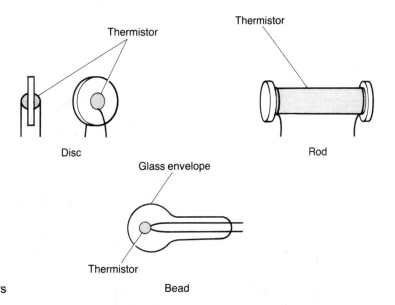

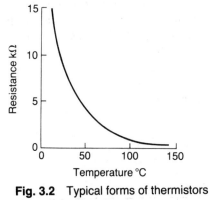

Fig. 3.2 Typical forms of thermistors

The resistance–temperature graph for a thermistor is highly non-linear and is described by the exponential relationship

$$R_t = Ke^{\beta/t}$$

where R_t is the resistance at temperature t, with K and β being constants.

The response time of a thermistor depends on the amount of thermistor material present. A small bead thermistor might have a response time of the order of $0.5\,s$ while large rod thermistors might be of the order of $10\,s$.

For thermistors which show a decrease in resistance with an increase in temperature, at low potential differences a small current is produced and the power dissipated is not sufficient to raise the temperature of the thermistor above that of its environment. However, at higher potential differences the current may be large enough for the power developed to be

Fig. 3.3 Resistance–temperature graph for a thermistor

sufficient to raise the temperature of the thermistor above that of its environment. An increase in temperature however means a decrease in resistance. As a result of this the current increases yet further. This results in a yet higher temperature and a greater decrease in resistance. The current then increases even more and results in yet higher temperatures. This effect continues until the heat dissipation of the thermistor equals the power supplied to it. This self heating effect is typically of the order of 0.15°C for every mW of supplied power.

A consequence of self heating is that if the surroundings are at a constant temperature the thermistor resistance depends on the rate at which heat is conducted away from it. It can thus be immersed in a moving fluid and used to measure the flow rate.

Strain gauges

The term *strain gauge* is used for a metal wire or foil element or a semiconductor strip which is wafer-like and can be stuck onto surfaces like a postage stamp and whose resistance changes when subject to strain (Fig. 3.4). The fractional change in resistance of the gauge, $\Delta R/R$, is directly proportional to the strain ϵ and so

$$\Delta R/R = G\epsilon$$

where G is a constant called the gauge factor.

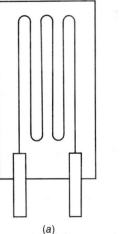
(a)

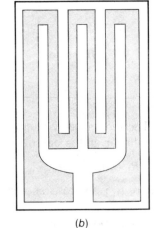
(b)

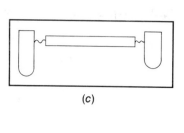
(c)

Fig. 3.4 Strain gauges, (a) metal wire, (b) metal foil, (c) semiconductor

The wire strain gauge consists of a length of wire wound in a grid shape and attached to a suitable backing material (Fig. 3.4(a)). Typical values of gauge factor for wire strain gauges which tend to use a copper-nickel-manganese alloy called Advance is about 2.0. This alloy has a low temperature coefficient of resistance and a low coefficient of linear

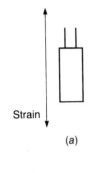

(a)

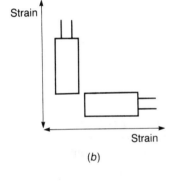

(b)

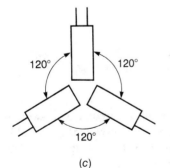

(c)

Fig. 3.5 Strain gauges with (a) uniaxial strain, (b) biaxial strain, (c) multidirectional strain

expansion. These are both desirable features if temperature effects are to be kept small in relation to the effects of strain on the gauge.

The metal foil strain gauge consists of a grid form which has been etched from a metal foil (Fig. 3.4(b)) and mounted on a resin film base. Typical values of gauge factor for metal foil gauges is about 2.0.

Semiconductor strain gauges (Fig. 3.4(c)) are generally just strips of silicon doped with small amounts of p or n-type material. They have very high gauge factors, of the order of 100 to 175 for p-type silicon and −100 to −140 for n-type silicon. The negative gauge factor means that the resistance decreases with an increase in strain, unlike all the other forms of gauges where a positive gauge factor indicates an increase in resistance with an increase in strain. Semiconductors with this great advantage over metal gauges of a high gauge factor do however have the disadvantage of a much greater temperature coefficient of resistance.

With a uniaxial strain a strain gauge can be mounted so that its axis is along the direction of the strain (Fig. 3.5(a)). If however there is biaxial strain a pair of gauges should be used, a gauge being mounted in each of the strain directions (Fig. 3.5(b)). Where the strain axes are not known or the strain is multidirectional a strain gauge rosette is used (Fig. 3.5(c)).

In order to compensate for the effects of temperature changes, for each active gauge a dummy gauge is generally used. The active and dummy gauges are mounted close together so that they are both at the same temperature. The active gauge is subject to the strain but the dummy gauge is not. By including the strain gauges in different arms of a Wheatstone bridge the effects of temperature on the resistance can be eliminated, leaving just the effect of the strain on the active gauge. See Chapter 4 for discussion of this technique with a Wheatstone bridge.

In some situations, such as bending, the effect of temperature changes can be eliminated by using two active gauges, one being on the compression side of the bent object the other on the tension side. When these are in different arms of the Wheatstone bridge the temperature effects cancel and the effect of the strain is doubled since the compression results in a decrease in resistance and the tension an increase in resistance.

Strain gauges, as well as being used for the measurement of strain, are widely used as transducers in other measurement systems. For example they can be used to sense the deformation of a diaphragm under the action of pressure and so give a resistance change related to pressure (Fig. 3.6). They can be used in what is called a *load cell* to give a resistance change related to force or weight. A load cell could be a cylinder with strain gauges mounted on its surfaces and the load applied along its axis (Fig. 3.7).

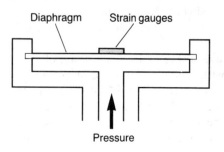

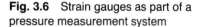

Fig. 3.6 Strain gauges as part of a pressure measurement system

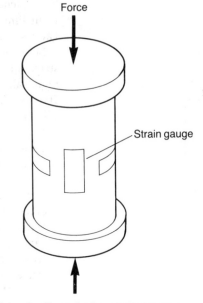

Fig. 3.7 A load cell employing strain gauges

Example 2

A strain gauge has a resistance of $120\,\Omega$ and a gauge factor of 2.1. What will be the change in resistance produced if the gauge is mounted with its axis in the direction of a uniaxial strain of 0.0005?

Answer

Using the equation given above

$$\Delta R/R = G\epsilon$$

Change in resistance $= \Delta R = RG\epsilon = 120 \times 2.1 \times 0.0005$
$$= 0.13\,\Omega$$

Potentiometers

The *rotary potentiometer* consists of a circular wire-wound track or a film of conductive plastic over which a rotatable electrical contact, the slider, can be rotated (Fig. 3.8). The track may be just a single turn or helical.

With a constant input voltage, between terminals 1 and 3 in Fig. 3.8, the output voltage V_o between the slider terminal 2 and terminal 1 depends on the position around the track to which the slider has been rotated. If the track has a constant resistance per unit length then the output is proportional to

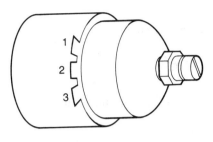

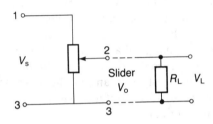

Fig. 3.8 Rotary potentiometer

the angle θ through which the slider has been rotated.

V_o is proportional to θ

$$V_o = k\theta$$

where k is a constant. Hence an angular displacement can be converted into a potential difference.

An important effect that has to be considered with a potentiometer is the effect of a load connected across the output, i.e. between terminals 2 and 3. By the use of a Thévenin equivalent circuit the potentiometer with a load R_L gives an output potential difference across the load of V_L where

$$V_L = V_s x \frac{1}{(R_p/R_L)x(1-x)+1}$$

where V_s is the supply voltage, and x is the fraction of the potentiometer resistance R_p between terminals 2 and 3. If R_L is infinity this equation becomes the linear relationship between V_L and x of $V_L = V_s x$. However for finite loads the result is a non-linear relationship and hence a non-linearity error. The error at a particular slider position of x is

$$\text{error} = V_s x - V_L$$

$$= V_s x \left[1 - \frac{1}{(R_p/R_L)x(1-x)+1} \right]$$

If R_p/R_L is considerably less than 1, which usually is the case, then the expression approximates to

$$\text{error} = V_s(R_p/R_L)(x^2 - x^3)$$

The resolution of a wire track is limited by the fineness of the wire used and typically ranges from about 1.5 mm for a coarsely wound track to 0.5 mm for a finely wound one. Errors due to non-linearity tend to range from less than 0.1% to about 1%. The track resistance tends to range from about 20 Ω to 200 kΩ. Conductive plastic has no resolution problems, errors due to non-linearity of the order of 0.05% and resistance values from about 500 Ω to 80 kΩ. The conductive plastic has a higher temperature coefficient of resistance than the wire and so temperature changes have a greater effect on the accuracy.

Example 3

What is the non-linearity error with a potentiometer of resistance 500 Ω which results from there being a load of resistance 10 kΩ when it is at a displacement which is half its maximum slider travel and the supply voltage is 4 V?

Answer

Using the approximate expression derived above

$$\text{error} = V_s(R_p/R_L)(x^2 - x^3)$$
$$= 4(500/10000)(0.5^2 - 0.5^3)$$
$$= 0.025\,\text{V}$$

As a percentage of the full-scale reading of V_s the error is

$$\text{error} = \frac{0.025}{4} \times 100 = 0.625\%$$

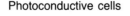
Photoconductive cells

A photoconductive cell has a resistance which depends on the intensity of light falling on it. Cadmium sulphide is a commonly used cell because it has a response to the colours of the spectrum which is very similar to that of the human eye. Such a cell can be used in a light meter or to detect the pulses of light produced as a result of a beam of light being interrupted by a rotating disc which contains windows (see rotating discs, later in this chapter).

Capacitive transducers

The capacitance C of a parallel-plate capacitor is given by

$$C = \frac{\epsilon_r \epsilon_o A}{d}$$

where ϵ_r is the relative permittivity of the dielectric between the capacitor plates, ϵ_o the permittivity of free space ($8.85 \times 10^{-12}\,\text{F/m}$), A the area of overlap between the two plates and d the plate separation.

Variable plate separation transducer

The capacitance of a parallel plate capacitor depends on the plate separation and so a change in this separation produces a change in capacitance. If the separation increases by x, as in Fig. 3.9, then the capacitance becomes

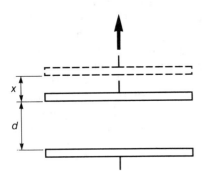

$$C = \frac{\epsilon_r \epsilon_o A}{d + x}$$

This is a non-linear relationship between the capacitance and the displacement x.

A pressure gauge based on this consists of a circular diaphragm, held at the edges, acting as one plate of the capacitor and a fixed plate for the other one. Changes in pressure cause the diaphragm to distort and so change the separation between it and the fixed plate. The result is a change in capacitance.

Fig. 3.9 Variable plate separation capacitive transducer

Variable plate area transducer

The capacitance of a parallel-plate capacitor depends on the overlap area of the two plates and so a change in area produces a change in capacitance. Figure 3.10 shows the form of a transducer based on this principle.

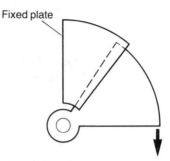

Fixed plate

Fig. 3.10 Variable plate area capacitive transducer

Variable dielectric transducer

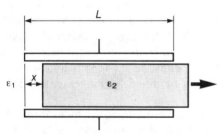

Fig. 3.11 Variable dielectric capacitive transducer

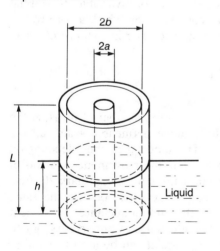

Fig. 3.12 Capacitive liquid level gauge

The capacitance of a parallel plate capacitor depends on the dielectric between the plates. Thus if the relative amount of two dielectrics between the plates varies then the capacitance varies. Figure 3.11 shows the basic form of such a transducer. The capacitance of such a capacitor is the sum of the capacitance of the two capacitors formed by the two dielectrics (it is the sum because the two capacitors are in parallel). Thus if the width of the plates is w then the capacitance C is

$$C = \frac{\epsilon_1\epsilon_0 wx}{d} + \frac{\epsilon_2\epsilon_0 w(L-x)}{d}$$

$$= \frac{\epsilon_0 w}{d}[\epsilon_2 L - (\epsilon_2 - \epsilon_1)x]$$

Such a transducer can be used for displacement measurement. One version is however used for liquid level measurement (Fig. 3.12). The arrangement consists of two concentric conducting cylinders. These form the plates of the capacitor. The capacitance per unit length C of coaxial cylinders, radii a and b, is given by

$$C = \frac{2\pi\epsilon_r\epsilon_0}{\ln(b/a)}$$

where ϵ_r is the relative permittivity of the medium between the cylinders. For the liquid level gauge there is the capacitance for that part of the cylinders between which there is the liquid and the capacitance of that part between which there is air. The two capacitors are in parallel and so the total capacitance is the sum of the two separate capacitances, one being of length h and the other $(L - h)$. Hence

$$C = \frac{2\epsilon_r\epsilon_o h}{\ln(b/a)} + \frac{2\epsilon_o(L - h)}{\ln(b/a)}$$

$$C = \frac{2\pi\epsilon_o}{\ln(b/a)}[L + (\epsilon_r - 1)h]$$

Inductive transducers

Variable reluctance

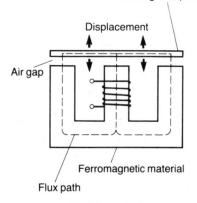

Ferromagnetic plate

Displacement

Air gap

Ferromagnetic material

Flux path

Fig. 3.13 Variable reluctance transducer

Figure 3.13 shows the basis of a variable reluctance transducer. The reluctance S of a magnetic circuit is given by

$$S = \frac{L}{\mu_r\mu_o A}$$

where μ_r is the relative permeability, μ_o is the permeability of free space ($4\pi \times 10^{-7}$ H/m), L the length of the magnetic flux path in the circuit and A the cross-sectional area of the flux path. The relative permeability of air is close to 1, while that of the ferromagnetic material is many thousands. This means that that part of the flux path through air is through a much higher reluctance path than that part through the ferromagnetic material. Reluctance is rather like resistance in an electrical circuit with the flux being the electrical current and the air being a high resistance while the ferromagnetic material is like the copper connecting wires and of low resistance. The reluctance of the air path depends on its length and in the displacement transducer this is modified by movement of the ferromagnetic plate.

Such a form of transducer can be used for the measurement of displacement and force, since the force can be used to produce a displacement of the ferromagnetic plate.

Variable differential inductor

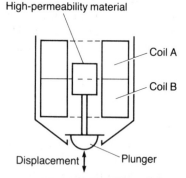

High-permeability material

Coil A

Coil B

Displacement   Plunger

Fig. 3.14 Variable differential inductor

Figure 3.14 shows the basic elements of a variable differential inductor. It consists of two coils between which a core rod of high permeability material is moved. The inductance of a coil is given by

$$\text{inductance} = \frac{\mu_r\mu_o N^2 A}{l}$$

where l is the length of the coil, N the number of turns, A the cross-sectional area, μ_r the relative permeability and μ_o the permeability of free space. The inductance thus depends on what material is in its core. Thus the movement of a rod of high permeability material into the core has a marked effect on the inductance of the coil. When the rod has the same length in each coil they have the same inductance. Movement of the rod then results in the inductance of one increasing and the other decreasing. A bridge circuit can be used to monitor this differential inductance which is a measure of the displacement of the rod.

Variable differential transformer

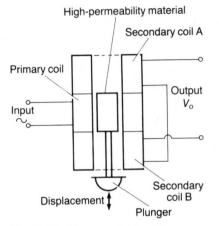

Fig. 3.15 Linear variable differential transformer

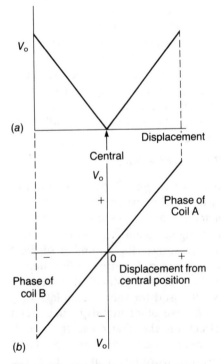

Fig. 3.16 LVDT characteristics, (a) a.c. characteristic, (b) d.c. characteristic

The linear variable differential transformer, generally referred to by the abbreviation LVDT, consists of a transformer with a primary coil and two secondary coils. Figure 3.15 shows the type of arrangement. The two secondary windings are connected in series so that their outputs oppose each other.

When there is an alternating voltage input to the primary coil, alternating e.m.f.s are induced in the secondary coils. With the high permeability core central, and since both the secondary coils are identical, the e.m.f.s induced in the two secondary coils will be the same. Since they are so connected that their outputs oppose each other, the result is zero output. However, when the core is displaced from the central position there is a greater amount in one coil than the other. The result is a change in the degree of coupling between the primary and the secondary coils and so the induced e.m.f. in one coil is greater than that in the other. The output, the difference between the two e.m.f.s, thus increases the more the core is moved away from its central position.

Figure 3.16(a) shows how the output V_o changes with the position of the core. This graph can be called the a.c. characteristic since the output is an alternating voltage. The zero to the central displacement part of the graph is when the core is in the secondary coil A and the primary coil, but not in B. The central displacement onwards is when the core is in the secondary coil B and the primary coil, but not in A.

For a sinusoidal input voltage the e.m.f.s in the two secondary coils can be represented by

$$V_A = k_A \sin(\omega t - \phi)$$

$$V_B = k_B \sin(\omega t - \phi)$$

where k_A and k_B depend on the degree of coupling occurring between the primary and secondary coils at some particular position of the core. Their values will change when the position of the core changes. ϕ is the phase difference between the primary alternating voltage and the secondary voltages. The output V_o is the difference between the two secondary voltages. Thus

$$V_o = V_A - V_B = (k_A - k_B) \sin(\omega t - \phi)$$

When the core is at its mid position k_A equals k_B and so V_o is zero. When, however, the core is at some position in the secondary coil A such that $k_A = k_1$ and $k_B = k_2$, then

$$V_o = (k_1 - k_2) \sin(\omega t - \phi)$$

If now the core is in the secondary coil B, by the same amount it was previously in the secondary coil A above, then coupling constants will be reversed with $k_A = k_2$ and $k_B = k_1$. Then

$$V_o = V_A - V_B = (k_A - k_B) \sin(\omega t - \phi)$$
$$= (k_2 - k_1) \sin(\omega t - \phi)$$
$$= (k_1 - k_2) \sin[\omega t + (\pi - \phi)]$$

The phase of the output differs from that occurring when the core was in secondary A by 180°.

The output voltage can be converted into d.c. in a way which distinguishes between the phases and so produce a d.c. characteristic of the form shown in Fig. 3.16(b).

Both the a.c and d.c form of the characteristics shown in Fig. 3.16 are ideal ones since there is inevitably some non-linearity near both extremes of core displacement. Non-linearity errors are likely to be between about 0.1% and 1%. The LVDT is used to measure displacements from about a quarter of a millimetre to 250 mm.

Thermoelectric transducers

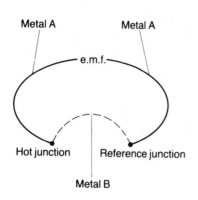

Fig. 3.17 A thermocouple

If two different metals are joined together a potential difference occurs across the junction. This potential difference depends on the metals used and the temperature of the junction. A thermocouple is a complete circuit involving two such junctions (Fig. 3.17). If both junctions are at the same temperature, then the e.m.f. produced by one junction cancels that from the other and there is no net e.m.f. If however there is a difference in temperature between the two junctions then there is an e.m.f., this being called the thermoelectric e.m.f. The value of this e.m.f. depends on the two metals concerned and the temperatures of both the junctions. Usually one junction is held at 0 °C and then

$$\text{e.m.f. } E = a_1 T + a_2 T^2 + a_3 T^3 + \text{etc.}$$

where a_1, a_2, a_3, etc. are constants and T the temperature. This is a non-linear relationship between the e.m.f. and temperature, however, for many pairs of metals the terms a_2, a_3, etc. are small enough to be ignored and a reasonably linear relationship occurs. Figure 3.18 shows, for a number of pairs of metals, how the thermoelectric e.m.f. varies with temperature when one junction is held at 0 °C. Standard tables are available for the metals usually used for thermocouples.

A thermocouple circuit can have other metals in the circuit and they will have no effect on the thermoelectric e.m.f. provided all their junctions are at the same temperature. Thus, for example, a voltage measuring instrument can be introduced into the circuit.

A thermocouple can be used with the reference junction at a temperature other than 0 °C. However the standard tables that are available for thermocouples assume a 0 °C junction and hence a correction has to be applied before the tables can be used. The correction is applied using what is known as the *law of intermediate temperatures*.

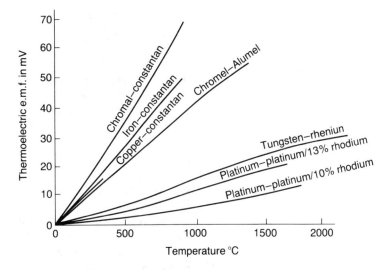

Fig. 3.18 Thermoelectric e.m.f.–temperature graphs

$$E_{T,0} = E_{T,I} + E_{I,0}$$

The e.m.f. at temperature T when the cold junction is at $0\,°C$, $E_{T,0}$, equals the e.m.f. at temperature T when the cold junction is at the intermediate temperature I, $E_{T,I}$, plus the e.m.f. at temperature I when the cold junction is at $0\,°C$, $E_{I,0}$.

To maintain one junction of a thermocouple at $0\,°C$, i.e. have it immersed in a mixture of ice and water, is often not very convenient. A compensation circuit can however be used instead. The circuit is required to generate a potential difference which will vary with the temperature of the cold junction in such a way that when it is added to the e.m.f. produced by the thermocouple it is as though the cold junction was at $0\,°C$. It has thus to supply the correction potential difference of $E_{I,0}$. Thus for an iron–constantan thermocouple this would mean a potential difference about $53\,\mu V$ for each degree the cold junction temperature is above $0\,°C$. Such a potential difference can be produced using a resistance thermometer (usually of nickel or copper wire) in a Wheatstone bridge (see Chapter 4). The bridge is balanced at $0\,°C$ so that there is no output potential difference at that temperature. Then the out-of-balance potential difference produced at other temperatures is the compensation potential difference.

Commonly used thermocouples are shown in Table 3.5 with the temperature ranges over which they are generally used and typical sensitivities. These commonly used thermocouples are given reference letters. For example the iron–constantan thermocouple is called a type J thermocouple.

The base-metal thermocouples, E, J, K and T, are relatively cheap but deteriorate with age. They have accuracies which typically are about ± 1 to 3%. The noble-metal thermocouples,

Table 3.5 Thermocouples

Type	Materials	Range °C	Sensitivity $\mu V/°C$
E	chromel–constantan	0 to 980	63
J	iron–constantan	– 180 to 760	53
K	chromel–alumel	– 180 to 1260	41
R	platinum–platinum/ rhodium 13%	0 to 1750	8
T	copper–constantan	– 180 to 370	43

e.g. type R, are more expensive but more stable with longer life. They have accuracies of the order of $\pm 1\%$ or less.

Thermocouples are generally mounted in a sheath to give them mechanical and chemical protection. The type of sheath used depends on the temperatures at which the thermocouple is to be used. In some cases the sheath is packed with a mineral which is a good conductor of heat and a good electrical insulator. The response time of an unsheathed thermocouple is very fast. With a sheath this may increase to as much as a few seconds if a large sheath is used. In some instances a group of thermocouples are connected in series so that there are perhaps ten or more hot junctions sensing the temperature. The e.m.f.s produced by each are added together. Such an arrangement is known as a *thermopile*.

Example 4

A type E thermocouple is to be used for the measurement of temperature with a cold junction at 20 °C. What will be the thermoelectric e.m.f. at 200 °C? The following is data from standard tables.

Temperature °C	0	20	200	180
e.m.f. mV	0	1.192	13.419	11.949

Answer

Using the law of intermediate temperatures

$$E_{200,0} = E_{200,20} + E_{20,0}$$
$$E_{200,20} = 13.419 - 1.192 = 12.227 \, mV$$

Note that the e.m.f. indicated for 180 °C, the difference between the hot and cold junction temperatures, is not the correct answer.

Example 5

An iron–constantan thermocouple is to be used to measure temperatures between 0 °C and 400 °C. What will be the non-linearity error as a percentage of the full-scale reading at 100 °C if a linear relationship is assumed between e.m.f. and temperature?
E.m.f. at 100 °C 5.268 mV, e.m.f. at 400 °C 21.846 mV

Answer

If there was a linear relationship then the e.m.f. at 100°C would be one quarter of that at 400°C, i.e. 5.4615 mV. This is the e.m.f. for which the assumed linear thermocouple indicates 100°C. The actual e.m.f. is 5.268 mV at 100°C and so there is an error of −0.1935 mV. As a percentage of the full-scale reading

$$\text{error} = -\frac{0.1935}{21.846} \times 100 = -0.89\%$$

Piezo-electric transducers

When a crystal is stretched or squashed, ions in the crystal are displaced from their normal positions. In some crystals this results in one face of the crystal becoming positively charged and the opposite face negatively charged, the amount of charge depending on the forces applied and the crystal concerned. This charge separation results in a potential difference appearing across the crystal. This effect is called *piezo-electricity*. Suitable crystals are quartz, tourmaline and what are called piezo-electric ceramics, such as lead zirconate-titanate. Piezo-electric transducers are widely used for the measurement of acceleration, vibration, and fluctuating pressures and forces.

The inverse piezo-electric effect is also possible, i.e. a potential difference applied across a crystal causes it to contract or expand. This effect is widely used for the production of ultrasonics.

Photovoltaic transducers

Junction diodes and transistors are normally enclosed in a screening can to prevent light affecting them. In the absence of such protection the output of the device is affected by light with a photocurrent being produced.

Elastic transducers

The spring balance is a widely used device for the measurement of forces which depends on forces producing changes in shape of some object, in this case a spring. A wide variety of such elastic transducers are used for the measurement of forces, pressures and torques.

Figure 3.19 shows a form of elastic transducer used for forces. It consists of a steel ring, called a *proving ring*, which deforms under the action of forces, the amount of deformation being a measure of the force. This can be measured by strain gauges (see earlier in this chapter) attached to the ring or a dial test indicator gauge as illustrated. Proving rings are capable of high accuracy and are used for forces in the range 2–2000 kN.

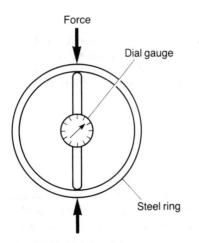

Fig. 3.19 Proving ring

Load cells, as described earlier in this chapter, employing strain gauges to monitor the contraction of some element, perhaps a cylinder, are widely used in industry for the monitoring of the weights of containers and hence a measure of the amount of material in them. Typically they are used for forces in the range 500 N to 6000 kN.

For the measurement of pressure the elastic deformation of diaphragms, capsules, bellows and tubes is widely used. Figure 3.20 shows some of the commonly used shapes. The amount of movement with a plane diaphragm (Fig. 3.20(a)) is fairly limited, however greater movement is possible with corrugations in the diaphragm (Fig. 3.20(b)). Even greater movement is possible if two corrugated diaphragms are combined to give a capsule (Fig. 3.20(c)). A stack of capsules is just a bellows arrangement (Fig. 3.20(d)) and this is even more sensitive.

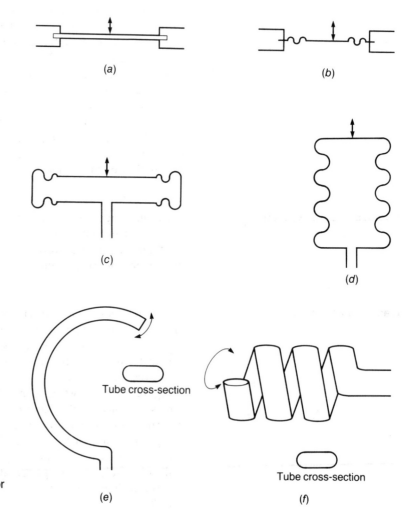

Fig. 3.20 Elastic transducers for pressure measurement

Diaphragms, capsules and bellows are made from such materials as stainless steel, phosphor bronze, and nickel, with rubber and nylon also being used for some diaphragms. Pressures in the range about 10^3–10^8 Pa can be measured by such transducers.

A different form of deformation is obtained using a C-shaped tube (Fig. 3.20(e)), this being generally known as a Bourdon tube. This opens up to some extent when the pressure in the tube increases. A helical form of such a tube (Fig. 3.20(f)) gives a greater deflection. The tubes are made from such materials as stainless steel and phosphor bronze and are used for pressures in the range 10^3–10^8 Pa.

Pneumatic transducers

Pneumatic transducers involve the use of compressed air, Fig. 3.21 showing a basic form. Air at a constant pressure P_s above the atmospheric pressure, i.e. the gauge pressure, flows through the orifice and escapes through the nozzle into the atmosphere. The pressure of the air between the orifice and nozzle is measured. The escape of air from the nozzle is controlled by the movement of the flapper. When this closes off the nozzle, i.e. $x = 0$, then no air escapes and the measured pressure equals P_s. As x increases so the pressure P decreases, becoming equal to the atmospheric pressure when x is very large. Figure 3.22 shows how the measured pressure P varies with the displacement x of the flapper (note that P is the gauge pressure, i.e. the pressure of the air above the atmospheric pressure). The measured pressure can thus be used as a measure of the displacement x.

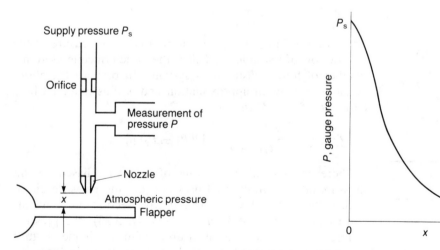

Fig. 3.21 Flapper-nozzle

Fig. 3.22 Flapper-nozzle characteristic

The relationship between the pressure P and the displacement x is given by

$$P = \frac{P_s}{1 + 16(d_n^2 x^2/d_o^4)}$$

where d_n is the diameter of the nozzle and d_o the diameter of the orifice. The relationship is non-linear. The transducer has high sensitivity but a small range of measurement, typically $\pm 0.05\,\text{mm}$.

Differential pressure transducers

When a fluid flows through a constriction in a pipe its velocity increases and the pressure drops. For flow which is incompressible, i.e. the density does not change when the pressure changes, Bernoulli's equation can be applied. For a horizontal pipe where v_1 is the fluid velocity, P_1 the pressure and A_1 the cross-sectional area of the pipe, v_2 the velocity, P_2 the pressure and A_2 the cross-sectional area at the constriction, and ρ the fluid density:

$$\frac{v_1^2}{2g} + \frac{P_1}{\rho g} = \frac{v_2^2}{2g} + \frac{P_2}{\rho g}$$

Since the density does not change, the volume of fluid Q passing through the wide section per second must equal the volume passing through the constriction (the equation of continuity). Hence

$$Q = A_1 v_1 = A_2 v_2$$

where A_1 is the cross-sectional area of the tube and A_2 that at the constriction. Hence

$$Q = \frac{A_2}{\sqrt{[1 - (A_2/A_1)^2]}} \sqrt{[2(P_1 - P_2)/\rho]}$$

Thus the pressure difference can be used as a measure of the rate of flow of the fluid Q. This is the basic principle used in a number of flow measurement systems. In practice the above equation is only an approximation and is thus modified by a correction factor C to

$$Q = \frac{CA_2}{\sqrt{[1 - (A_2/A_1)^2]}} \sqrt{[2(P_1 - P_2)/\rho]}$$

There are a number of forms of flowmeter based on this measurement of differential pressure, i.e. the pressure difference between the flow in the full cross-section tube and the constriction. With the *Venturi tube* (Fig. 3.23), the pressure difference is sometimes measured with a simple U-tube manometer connected between the two points, i.e. prior to the

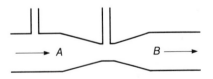

Fig. 3.23 Venturi tube

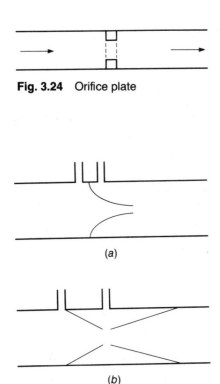

Fig. 3.24 Orifice plate

(a)

(b)

Fig. 3.25 (a) Nozzle flowmeter,
(b) Dall flowmeter

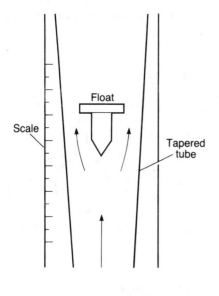

Scale

Float

Tapered
tube

Fig. 3.26 Variable-area flowmeter

constriction and at the constriction. The *orifice flowmeter* (Fig. 3.24) is simply a disc, with a central hole, which is placed in the tube. This results in a similar flow pattern to that occurring with the Venturi tube. The pressure difference is measured between a point equal to the diameter of the tube upstream and a point a distance equal to half the diameter downstream. Figure 3.25 shows two further versions of flowmeter, the *nozzle flowmeter* and the *Dall flowmeter*. The Venturi tube offers the least resistance to fluid flow and thus has the least effect on the rate of flow of fluid through the pipe while the orifice plate offers the greatest resistance. Typical values of the correction factor C are: Venturi 0.99, nozzle 0.96, Dall 0.66, orifice 0.60. All the flowmeters have long life without maintenance or recalibration being required and an accuracy of about $\pm 0.5\%$.

Another form of constriction flowmeter is the *variable area flowmeter*. This depend on the same principle but instead of measuring the pressure difference between the full-width tube and the constriction, the size of the constriction is varied to give a constant pressure difference. A common form of variable area flowmeter is the *rotameter* (Fig. 3.26). This employs a float in a tapered vertical tube. The fluid in flowing up the tube has to flow through the constriction which is the gap between the float and the walls of the tube. The result is a pressure difference which pushes the float up the vertical tube. The float stops moving up the tube when the pressure difference is just sufficient to balance the weight of the float. The tube is tapered and thus the gap between the float and the tube walls increases as the float moves up the tube. The greater the flow rate the greater the pressure difference for a particular gap. The float thus moves up the tube to a height which depends on the rate of flow. A scale alongside the tube can thus be calibrated to read directly the flow rate corresponding to a particular height of float. The rotameter can be used to measure flow rates from about 30 ml/s (30×10^{-6} m³/s) to 120 l/s (120×10^{-3} m³/s). It is a relatively cheap instrument, capable of a long life with little maintenance or recalibration being required. It is not highly accurate, about $\pm 1\%$, and has a significant effect on the flow.

Constriction flow meters measure the volume rate of flow. An instrument which can be used to measure the fluid velocity at some locality in a fluid flow is the *Pitot static tube* (Fig. 3.27). In this instrument the pressure difference is measured between a point in the fluid where the fluid is in full flow and so has both kinetic and potential energy and a point at rest in the fluid where there is only potential energy. The difference in pressure is thus due solely to the kinetic energy. Hence, for an incompressible fluid,

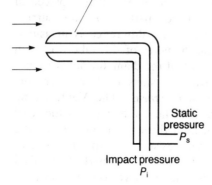

Holes around circumference of pipe

Static pressure P_s

Impact pressure P_i

Fig. 3.27 Pitotstatic tube

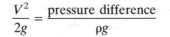

$$\frac{V^2}{2g} = \frac{\text{pressure difference}}{\rho g}$$

where ρ is the fluid density.

Pitot static tubes can be used for measurements of fluid velocities as low as 1 m/s and as high as 60 m/s and for both liquid and gas flow, though with gases the equation might need modification because the flow is compressible.

Example 6

A Venturi tube is to be used to measure a flow rate of water in a 100 mm diameter pipe. What diameter constriction would be suitable if the maximum flow rate is likely to be 0.20 m³/s and the maximum differential pressure to be measured 1.0 × 10⁵ Pa? Ignore any correction factor. The density of water is 1000 kg/m³.

Answer

Using the equation developed above,

$$Q = \frac{A_2}{\sqrt{[1 - (A_2/A_1)^2]}} \sqrt{[2(P_1 - P_2)/\rho]}$$

$$0.20 \times \sqrt{[1 - (A_2/A_1)^2]} = A_2 \times \sqrt{[2 \times 1.0 \times 10^5 \times 1000]}$$

Since $A_1 = \frac{1}{4}\pi \times 0.100^2$ then the diameter giving the cross-sectional area A_2 is 6.4 mm.

Turbine transducers

The turbine flowmeter (Fig. 3.28) consists of a multi-bladed rotor that is supported centrally in the pipe along which the flow occurs and which rotates as a result of the fluid flow. The angular velocity of the rotation is approximately proportional to the flow rate. The rate of revolution of the rotor can be

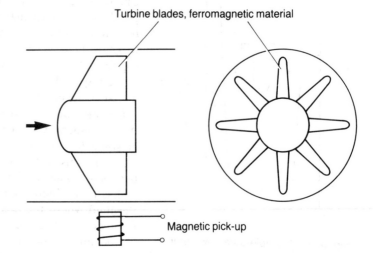

Turbine blades, ferromagnetic material

Magnetic pick-up

Fig. 3.28 Turbine meter

determined using a magnetic pick-up (see the next item on rotating ferromagnetic wheel) which produces an induced e.m.f. pulse every time a rotor blade passes it. The pulses are counted and so the number of revolutions of the rotor determined. The meter is used with liquids and offers some resistance to fluid flow. It is expensive. The accuracy is typically about $\pm 0.3\%$.

Rotating discs

Rotating ferromagnetic wheel

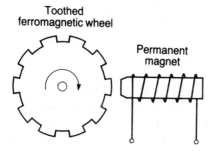

Fig. 3.29 Electromagnetic transducer

Shaft encoders

In the form used for the measurement of the angular speed of a rotating shaft, the arrangement consists of a toothed ferromagnetic wheel which rotates with the shaft and a detector to give a pulse each time a tooth passes it (Fig. 3.29). The detector consists of a permanent magnet around which a coil is wound. The arrangement constitutes a magnetic circuit which has a variable air gap, depending on whether a tooth or a gap is opposite the magnet. Changing the air gap in the magnetic circuit changes the circuit reluctance and hence the flux passing through the coil. The result of changing the flux linked by the coil is to induce an e.m.f. in it and so a series of pulses are produced as the teeth pass the end of the magnet. The pulses can be counted in a given time to give an output related to the angular velocity of the shaft.

The term *shaft encoders* is used for a device which gives an output in digital form related to the angular position of a shaft. The *incremental shaft encoder* gives just on-off signals as the shaft rotates and the angular position of the shaft can only be determined by counting the number of these pulses that has occurred since the shaft was at some particular position.

Figure 3.30 shows the form of a simple incremental shaft encoder. It consists of a disc which rotates along with the shaft. The form of the disc depends on the transducer used with it. In the optical form the disc has a series of windows through which a beam of light can pass. The beam of light falls on a light-sensitive transducer which gives an electrical output. Rotation of the shaft means that the light beam is pulsed and so the electrical output from the transducer is a series of pulses. These can be counted and so the angular position of the shaft determined by the number of pulses produced since some datum position. Where the rate of rotation of the shaft is required then the number of pulses produced per second is determined.

The *digital shaft encoder* gives an output in the form of a binary number of several digits, the number depending on the angular position of the shaft, and this provides an absolute measurement of the angular position of the shaft. Each angular position of the shaft has its own binary code. Figure

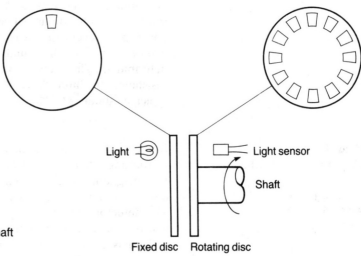

Fig. 3.30 An optical incremental shaft encoder

3.31 shows the form of the device. With the optical form the rotating disc has four concentric circles of slots and four sensors to detect the light pulses. The slots are arranged in such a way that the sequential output from the sensors is a number in the binary code. The result is that the output from the sensors indicates the angular position of the shaft. The disc shown in Fig. 3.31 gives the conventional form of binary code. Other forms of code are also used.

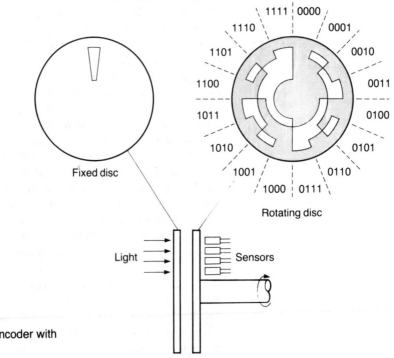

Fig. 3.31 Optical shaft encoder with coded disc

Other forms of detecting the pulses produced by the rotating of the discs are used. Thus with an electrical method the disc is made of insulating material and has a metallic pattern printed on it. A d.c. voltage is connected to the metallic pattern by means of a brush and the segments of metal can be detected by means of brushes. The electromagnetic form involves a metallic disc with the pattern produced by regions of high and low permeability material. The sensors used are then magnets with coils wound on them, of the form described earlier under the heading rotating ferromagnetic disc.

Problems

1 Transducers change information about a quantity into information in another form. For the following transducers, what information changes are occurring?
(a) Thermocouple.
(b) Photoconductive cell.
(c) Strain gauge.
(d) Linear variable differential transformer.
(e) Venturi gauge.

2 Specify a transducer, or primary and secondary transducers, which could be used for each of the following information change situations.
(a) Displacement to potential difference.
(b) Force to displacement.
(c) Force to resistance change.
(d) Angular position of a shaft to pulses of light.
(e) Temperature to resistance.

3 A copper–constantan thermocouple is to be used to measure temperatures between 0 °C and 200 °C. What will be the non-linearity error as a percentage of the full-scale reading at 100 °C if a linear relationship is assumed between e.m.f. and temperature? E.m.f. at 100 °C 4.277 mV, e.m.f. at 200 °C 9.286 mV.

4 A platinum resistance thermometer has a resistance of 100.00 Ω at 0 °C, 138.50 Ω at 100 °C and 175.83 Ω at 200 °C. What will be the non-linearity error in °C at 100 °C if the thermometer is assumed to have a linear relationship between 0 °C and 200 °C?

5 A manufacturer of alarm systems is investigating the possibility of producing a detector which will cause a light to flash when the temperature in a room, e.g. a cold store for meat or other produce, reaches a particular temperature. What characteristics will be required of the transducer? Suggest possible transducers with a justification for your suggestions.

6 What will be the change in resistance of an electrical resistance strain gauge with a gauge factor of 2.1 and resistance 100 Ω if it is used to measure a strain of 0.1%?

7 What is the non-linearity error, as a percentage of full-scale reading, produced when a 1 kΩ potentiometer has a load of 10 kΩ and is at one-third of its maximum displacement?

8 A Venturi tube is to be used to measure a flow rate of water in a 500 mm diameter pipe. What diameter constriction would be

suitable if the maximum flow rate is likely to be $0.80\,m^3/s$ and the maximum differential pressure to be measured $3.0 \times 10^5\,Pa$? Ignore any correction factor. The density of water is $1000\,kg/m^3$.

9 For safety reasons it is important that the load carried in a hoist or lift, whether passengers or objects, does not exceed the permitted amount for the hoist/lift cables. Consider what the requirements are for the transducers needed to monitor the weight carried and propose possible solutions.

10 The following are typical extracts from specifications of transducers. Interpret them and explain the circumstances under which the transducer would be most usefully used.

(a) A strain gauge pressure transducer
Range 2–70 MPa, 70 kPa to 1 MPa
Excitation 10 V d.c. or a.c. (r.m.s.)
Full-range output 40 mV
Non-linearity and hysteresis ±0.5%
Temperature range −54 °C to +120 °C
Thermal shift zero: 0.030% full-range output/°C
Thermal shift sensitivity: 0.030% full-range output/°C.

(b) A load cell
Ranges available 0–50 kg, 0–100 kg, 0–200 kg, 0–500 kg, 0–1000 kg, 0–2000 kg, 0–5000 kg
Total error due to non-linearity, hysteresis and non-repeatability ±0.25%
Application: low–medium accuracy container weighing.

(c) A thermistor
Resistance 2000 Ω at 20 °C, 40 Ω at 200 °C
Temperature range −100 to +250 °C
Self-heating effect 1 °C rise in temperature per 1.4 mW power dissipation.

(d) Electrical resistance strain gauge
Gauge resistance 100 Ω ±0.5%
Gauge factor 2.1 ±1%
Temperature range −20 °C to +60 °C
Temperature coefficient of resistance $<0.05\,°C^{-1}$
Coefficient of expansion $1.1 \times 10^{-7}°C^{-1}$.

(e) Turbine flowmeter
Range 1–15 l/s
Accuracy ±0.5%
Head loss 30 kPa
Response time 15 μs
Pulse rate 160 pulses/l.

4 Signal conditioning and processing

Introduction

The term *signal conditioning* is used for the element or elements in a measurement or control circuit which convert the signal from the transducer into a form suitable for further processing. In a measurement system this could mean a form suitable for the display unit. Typical signal conditioning elements are bridges where a change in resistance, capacitance or inductance can be converted to a change in potential difference.

The term *signal processing* is used for the element or elements which are concerned with improving the quality of the signal. This involves such processes as signal amplification, signal attenuation, signal linearization, and signal filtering.

Wheatstone bridge – null method

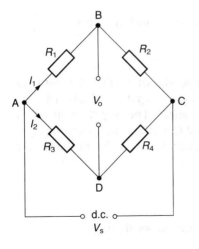

Fig. 4.1 The Wheatstone bridge

Figure 4.1 shows the basic form of the *Wheatstone bridge*. It has a d.c. supply and each of the four bridge arms is a resistance. When used as a null-method of signal conditioning the resistances in the arms of the bridge are so adjusted that the output potential difference is zero. If a galvanometer is connected between the output terminals this means the resistances are adjusted to give zero current through it. In such a condition the bridge is said to be balanced.

When the output potential difference is zero then the potential at B must equal that at D. This means that the potential difference across R_1, i.e. V_{AB}, must equal that across R_3, i.e. V_{AD}. Thus

$$I_1 R_1 = I_2 R_3$$

It also means that the potential difference across R_2, i.e. V_{BC}, must equal that across R_4, i.e. V_{DC}. Since there is no current through BD then the current through R_2 must be I_1 and that through R_4 I_2. Thus

$$I_1 R_2 = I_2 R_4$$

Hence

$$I_1R_1 = I_2R_3 = (I_1R_2/R_4)R_3$$

$$\frac{R_1}{R_2} = \frac{R_3}{R_4}$$

The balance condition is independent of the supply voltage, depending only on the resistances in the four bridge arms. If R_2 and R_4 are known fixed resistances and R_1 the unknown or output from the transducer then R_3 can be adjusted to give the zero potential difference condition and R_1 determined from a knowledge of the values of R_2, R_3 and R_4. By a suitable choice of the ratio R_2/R_4 a small resistance change in R_1 can be determined by means of a much larger resistance change in R_3.

Example 1

A Wheatstone bridge has a resistance ratio of 1/100 for R_2/R_4 and R_3 is adjusted to give zero current. Initially this occurs with R_3 1000.3 Ω. The resistance R_1 then changes as a result of a temperature change and zero current is then obtained when R_3 is 1002.1 Ω. What was the change in resistance of R_1?

Answer

Initially

$$R_1 = \frac{R_2}{R_4}R_3 = \frac{1}{100} \times 1000.3$$

After the change

$$R_1 + \text{change in } R_1 = \frac{1}{100} \times 1002.1$$

$$\text{change in } R_1 = \frac{1}{100} \times (1002.1 - 1000.3) = 0.018\,\Omega$$

Thus a change of 0.018 Ω in R_1 is determined by a change in R_3 of 1.8 Ω.

Wheatstone bridge – deflection type Consider the Wheatstone bridge shown in Fig. 4.1 with no galvanometer connected across the output terminals, i.e. the output load has infinite resistance. The supply voltage is connected between points A and C and thus the potential drop across the resistor R_1 is the fraction $R_1/(R_1 + R_2)$ of the supply voltage V_s. Hence

$$V_{AB} = \frac{V_sR_1}{R_1 + R_2}$$

Similarly, the potential difference across R_3 is

$$V_{AD} = \frac{V_s R_3}{R_3 + R_4}$$

Thus the difference in potential between points B and D, i.e. the output potential difference V_o, is

$$V_o = V_{AB} - V_{AD} = V_s \left(\frac{R_1}{R_1 + R_2} - \frac{R_3}{R_3 + R_4} \right)$$

This equation gives the null condition equation when V_o is equated to zero.

If the transducer is the resistor R_1 then the relationship between the output potential difference V_o and its resistance R_1 is a non-linear relationship.

A change in resistance from R_1 to $R_1 + \delta R_1$ gives a change in output from V_o to $V_o + \delta V_o$, where

$$(V_o + \delta V_o) - V_o = V_s \left(\frac{R_1 + \delta R_1}{R_1 + \delta R_1 + R_2} - \frac{R_1}{R_1 + R_2} \right)$$

If δR_1 is much smaller than R_1, which is frequently the case, then the equation approximates to

$$\delta V_o = \frac{V_s \delta R_1}{R_1 + R_2}$$

Under such conditions the change in output potential difference is proportional to the change in resistance of the transducer.

The above analysis is concerned with the open-circuit voltage between B and D in Fig. 4.1. If however there is a galvanometer of resistance R_G between the two points then a current I_G is drawn. This current can be determined with the Thévenin equivalent circuit (Fig. 4.2). The Thévenin voltage V_{Th} is the open circuit voltage V_o derived above. Thus

$$V_{Th} = V_s \left(\frac{R_1}{R_1 + R_2} - \frac{R_3}{R_3 + R_4} \right)$$

The Thévenin resistance R_{Th} is the resistance seen at points B and D of the bridge and is, if the d.c. supply has negligible internal resistance,

$$R_{Th} = \frac{R_1 R_2}{R_1 + R_2} + \frac{R_3 R_4}{R_3 + R_4}$$

The current I_G is thus

$$I_G = \frac{V_{Th}}{R_{Th} + R_G}$$

The potential difference across the galvanometer V_G is

$$V_G = I_G R_G$$

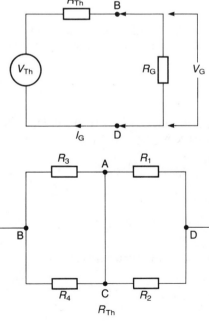

Fig. 4.2 Thévenin equivalent circuit

Example 2

A platinum resistance thermometer has a resistance at $0\,°C$ of $100\,\Omega$ and forms one arm of a Wheatstone bridge. At this temperature the bridge is balanced with each of the other arms also being $100\,\Omega$. The temperature coefficient of resistance of the platinum is $0.0039\,°C^{-1}$. What will be the output voltage per $°C$ change in temperature if the instrument used to measure it can be assumed to have infinite resistance and the supply voltage, with negligible internal resistance, for the bridge is $6.0\,V$?

Answer

The resistance variation of the platinum resistance thermometer can be represented by (see Ch. 3)

$$R_t = R_o(1 + \alpha t)$$

where R_t is the resistance at temperature t, R_o it at $0\,°C$ and α the temperature coefficient of resistance. Thus

$$\text{change in resistance} = R_t - R_o = R_o\alpha t$$

The change in resistance for a temperature change of $1\,°C$ is thus

$$\text{change in resistance} = 100 \times 0.0039 \times 1 = 0.39\,\Omega$$

Since this resistance is small compared with the $100\,\Omega$ the approximate equation derived above for the open circuit voltage can be used. Thus

$$\delta V_o = \frac{V_s\delta R_1}{R_1 + R_2} = \frac{6.0 \times 0.39}{100 + 100} = 0.012\,V$$

Example 3

For the Wheatstone bridge shown in Fig. 4.3 what will be the out-of-balance current through the galvanometer? The d.c. supply may be assumed to have negligible resistance.

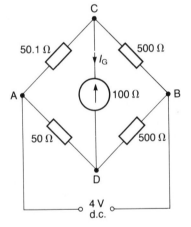

Fig. 4.3 Example 3

Answer

The Thévenin equivalent circuit for the bridge is shown in Fig. 4.4. The Thévenin equivalent resistance is given by

$$R_{Th} = \left(\frac{500 \times 50.1}{500 + 50.1} + \frac{500 \times 50}{500 + 50}\right) = 99.99\,\Omega$$

The Thévenin voltage is given by

$$V_{Th} = \left(\frac{50.1}{50.1 + 500} - \frac{50}{50 + 500}\right) = 6.61 \times 10^{-4}\,V$$

Hence the current I_G through the galvanometer is given by

$$I_G = \frac{V_{Th}}{R_{Th} + 100} = \frac{6.61 \times 10^{-4}}{99.99 + 100} = 3.31 \times 10^{-6}\,A$$

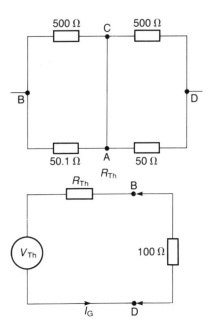

Fig. 4.4 Example 3

Wheatstone bridge – thermocouple compensation

A thermocouple gives an e.m.f. which depends on the temperatures of its two junctions (see Ch. 3). Usually one junction is kept at 0 °C. However this is not always feasible and the cold junction is often allowed to be at the ambient temperature. To compensate for this a potential difference has to be added to the thermocouple. This must be the same as the e.m.f. that would be generated by the thermocouple with one junction at 0 °C and the other at the ambient temperature. Such a potential difference can be produced by using a resistance thermometer in a Wheatstone bridge. The bridge is balanced at 0 °C and the out-of-balance potential difference provides the correction potential difference at other temperatures.

The resistance of a resistance thermometer element varies with temperature in a way that can be described by the relationship (see Ch. 3)

$$R_t = R_o(1 + \alpha t)$$

where R_t is the resistance at temperature t, R_o the resistance at 0 °C and α the temperature coefficient of resistance. Hence the change in resistance is

change in resistance $= R_t - R_o = R_o \alpha t$

Using the equation given earlier for the out-of-balance potential difference

$$\delta V_o = \frac{V_s \delta R_1}{R_1 + R_2}$$

and taking R_1 to be the resistance of the resistance thermo-meter element, then the out-of-balance potential difference δV_o is

$$\delta V_o = \frac{V_s R_o \alpha t}{R_o + R_2}$$

If the thermocouple e.m.f. e variation with temperature t can be represented by

$$e = at$$

where a is a constant (the e.m.f. produced per degree change in temperature). Then for compensation we must have

$$at = \frac{V_s R_o \alpha t}{R_o + R_2}$$

$$aR_2 = R_o(V_s \alpha - a)$$

Thus for an iron-constantan thermocouple giving $51\,\mu V/°C$, a nickel resistance element with a resistance of $10\,\Omega$ at $0°C$ and a temperature coefficient of resistance of $0.0067/°C$, and a supply voltage of $1.0\,V$, R_2 will need to have a value of $1304\,\Omega$.

Wheatstone bridge – compensation

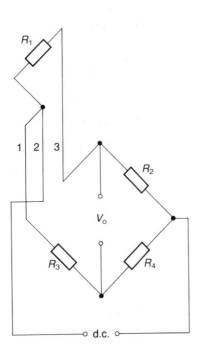

Fig. 4.5 Compensation for leads

In many measurement systems involving a resistive transducer the actual resistance element may have to be at the end of long leads. The resistance of these leads will be affected by changes in temperature. For example, a platinum resistance thermo-meter consists of a platinum coil at the ends of leads. When the temperature changes not only will the resistance of the platinum coil change but so also will the resistance of the leads. What is required is just the resistance of the coil and so some means has to be employed to compensate for the lead resistance. One method of doing this is to use three leads to the coil, as shown in Fig. 4.5. The coil is then connected into a Wheatstone bridge in such a way that lead 1 is in series with the R_3 resistor while lead 3 is in series with the platinum resistance coil R_1. Lead 2 is the connection to the power supply. Any change in lead resistance as a result of a temperature change will affect all three leads equally since they are all the same length and resistance. The result is that changes in lead resistance occur equally in two arms of the bridge and will cancel out if R_1 and R_3 are the same resistance.

The electrical resistance strain gauge is another transducer where compensation has to be made for temperature effects. The strain gauge changes resistance as the strain applied to it changes. Unfortunately it also changes resistance if the temperature changes. One way of eliminating the temperature effect is to use a dummy gauge. This is a strain gauge which is identical to the one under strain, the active gauge, but it is not subject to the strain. It is however at the same temperature as the active gauge. Thus a temperature change will cause both gauges to change resistance by the same amount. The active gauge is mounted in one arm of a Wheatstone bridge and the dummy gauge in another arm such that the effects of temperature-induced resistance changes cancel out (Fig. 4.6).

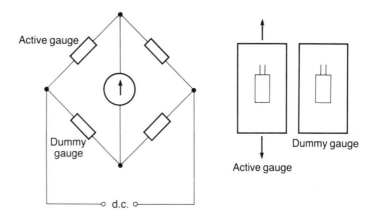

Fig. 4.6 Compensation with strain gauges

Strain gauges are often used as a secondary transducer, i.e. a means of converting the output of the initial transducer into a more convenient form. For example, a load cell suffers elastic deformation as the result of forces and this elastic deformation can be converted into a resistance change by strain gauges attached to the load cell (see Ch. 3). In such a situation temperature effects on the strain gauges have to be compensated for. While dummy gauges could be used, a better solution is to use four strain gauges. Two of them are attached to the load cell in such a way that they are in tension and two so that they are in compression when the forces are applied. The gauges that are in tension will increase in resistance while those in compression will decrease in resistance as a result of the application of the forces. If the gauges are connected as the four arms of a Wheatstone bridge, as in Fig. 4.7, then since all will be equally affected by any temperature changes the arrangement is temperature compensated. The arrangement also gives a much greater output voltage or galvanometer current than using just a single active gauge.

Alternating-current bridges

The a.c. bridge circuit (Fig. 4.8) is similar to the d.c. Wheatstone bridge with the same forms of relationships for balance and out-of-balance current. The basic conditions for balance, i.e. zero potential difference between A and D and hence zero current through the detector, are the potential

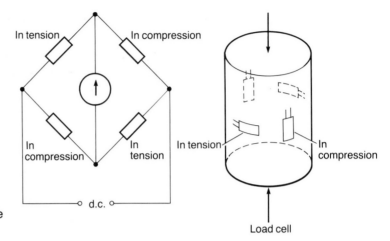

Fig. 4.7 Four active arm strain gauge bridge

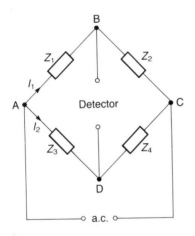

Fig. 4.8 Basic a.c. bridge

difference across Z_1 must be the same as that across Z_3, in both magnitude and phase, and similarly the potential differences across Z_2 and Z_4 must be equal.

$$I_1 Z_1 = I_2 Z_3$$
$$I_1 Z_2 = I_2 Z_4$$

Hence

$$\frac{Z_1}{Z_2} = \frac{Z_3}{Z_4}$$

There are many variations of the basic a.c. bridge, Fig. 4.9 shows a few of the more common ones and the interpretation of the above conditions for balance for each bridge.

To illustrate the application of the above condition for balance, consider the Maxwell bridge (Fig. 4.9a).

$$\frac{1}{Z_1} = \frac{1}{R_1} + j\omega C_1$$

$$Z_1 = \frac{R_1}{1 + j\omega C_1 R_1}$$

$$Z_2 = R_2$$
$$Z_3 = R_3$$
$$Z_4 = R_4 + j\omega L_4$$

Hence using the balance equation,

$$\frac{Z_1}{Z_2} = \frac{Z_3}{Z_4}$$

$$Z_4 = \frac{Z_2 Z_3}{Z_1}$$

$$R_4 + j\omega L_4 = \frac{R_2 R_3 (1 + j\omega C_1 R_1)}{R_1}$$

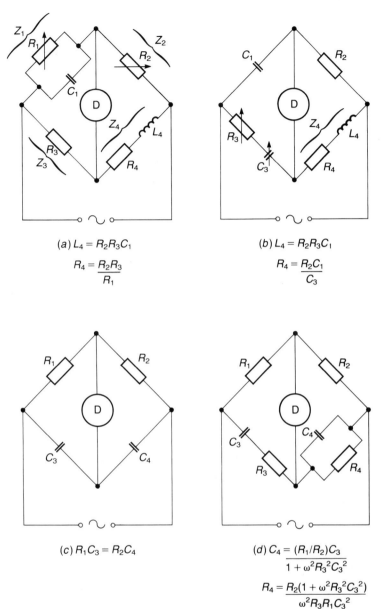

Fig. 4.9 Alternating-current bridges, (a) Maxwell, (b) Owen, (c) De Souty, (d) Wien

For the real parts we have a balance condition of

$$R_4 = \frac{R_2 R_3}{R_1}$$

For the imaginary parts

$$L_4 = R_2 R_3 C_1$$

Thus the resistance R_4 and the inductance L_4 of an inductor can be determined. The procedure is usually to adjust R_2 to obtain the best balance, then R_1 to improve it and then R_2 again and so on until a final balance is obtained.

Example 4

An a.c. bridge has in arm AB a $0.2\,\mu\text{F}$ capacitor, in arm BC a $500\,\Omega$ resistor, in arm CD a $0.1\,\text{H}$ inductor in series with a $50\,\Omega$ resistor and in arm DA an unknown inductor. The inductor can be considered to be a pure inductance in series with a pure resistance. What are the values of this inductance and resistance if the bridge is balanced at a frequency of $1\,\text{kHz}$?

Answer

The balance equation for an a.c. bridge is

$$\frac{Z_1}{Z_2} = \frac{Z_3}{Z_4}$$

If Z_1 is arm AB, Z_2 arm BC, Z_3 arm DA, Z_4 arm CD, then

$$Z_1 = \frac{1}{j\omega C_1}$$

$$Z_2 = R_2$$
$$Z_3 = R_3 + j\omega L_3$$
$$Z_4 = R_4 + j\omega L_4$$

Hence the balance equation becomes

$$\frac{1}{j\omega C_1 R_2} = \frac{R_3 + j\omega L_3}{R_4 + j\omega L_4}$$

$$R_4 + j\omega L_4 = j\omega C_1 R_3 R_2 - \omega^2 L_3 C_1 R_2$$

Thus for the real parts

$$R_4 = -\omega^2 L_3 C_1 R_2$$

and for the imaginary parts

$$\omega L_4 = \omega C_1 R_3 R_2$$

Hence

$$50 = -1000^2 \times L_3 \times 0.2 \times 10^{-6} \times 500$$

$$0.1 = 0.2 \times 10^{-6} \times R_3 \times 500$$

and so $L_3 = 0.5\,\text{H}$ and $R_3 = 1000\,\Omega$

Potentiometer measurement system The potentiometer measurement system involves using a potentiometer to produce a variable potential difference which can then be used to balance, and so cancel out, the potential difference being measured. Figure 4.10 shows the basis of the system. The working battery is used to produce a potential difference across the full length of the potentiometer track. The potential difference across the length L of the potentiometer track is then balanced against the unknown e.m.f., the length L being adjusted until no current is detected by the galvanometer. When this occurs the unknown e.m.f. E must be equal to the potential difference across the length L of the track. If the track is uniform then

$$E = kL$$

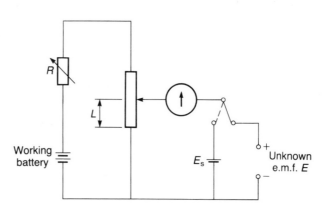

Fig. 4.10   Potentiometric measuring system

where k is a constant, in fact the potential difference per unit length of track. This can be determined by repeating the balancing operation with a standard cell of e.m.f. E_s. Then

$$E_s = kL_s$$

where L_s is the balance length with the standard cell. Hence

$$E = \frac{E_s}{L_s} L$$

With a commercial form of the potentiometer measuring system the movement of the potentiometer slider over the track results in the movement of a pointer over a scale. This scale is calibrated directly in volts. The standardisation is achieved by setting the pointer to the required value of the standard cell e.m.f. and then adjusting R until balance occurs.

The potentiometer measurement system does not depend on the calibration of a galvanometer since it is only used to indicate when there is zero current. Because at balance no current is taken from the source being measured, no power is taken from the source. This makes it a useful method for use with transducers which do not produce much power. The system is essentially an infinite impedance voltmeter.

Signal amplification

Amplifiers are frequently used as signal conditioners in order to make signals from transducers big enough to enable them to be further processed or displayed. Amplifiers can take various forms, e.g. mechanical, in the form of a lever or gear train, or electronic. Whatever the form a transfer function can be defined as

$$\text{transfer function} = \frac{\text{output}}{\text{input}}$$

Mechanical amplifier – the lever

The lever can be used to change the size of a displacement signal from a transducer. The transfer function of the lever depends on the relative distances from the lever pivot point of the application of the input to the lever and of the extraction of the output. As indicated in Fig. 4.11, because of similar triangles

$$\frac{\text{input displacement}}{\text{input–pivot distance}} = \frac{\text{output displacement}}{\text{output–pivot distance}}$$

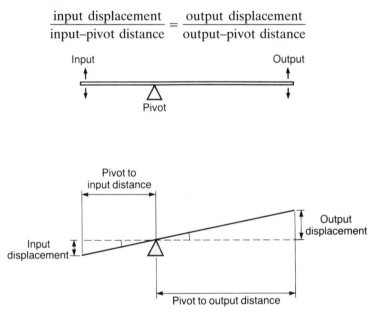

Fig. 4.11 The lever as a signal conditioner

Hence

$$\text{transfer function} = \frac{\text{output}}{\text{input}} = \frac{\text{output–pivot distance}}{\text{input–pivot distance}}$$

Where a large magnification is required a compound lever may be used. This involves the output from the first lever becoming the input to a second lever. Figure 4.12 shows an application of this in the Huggenberger extensometer.

Mechanical amplifier – gear trains

Figure 4.13 shows a simple gear train involving an input gear wheel which is rotated by one shaft and an output gear wheel which results in rotation of another shaft. Each tooth on the input gear wheel fits into a corresponding space between teeth on the output gear wheel. If there are N_I teeth on the input wheel and N_o on the output wheel then one complete

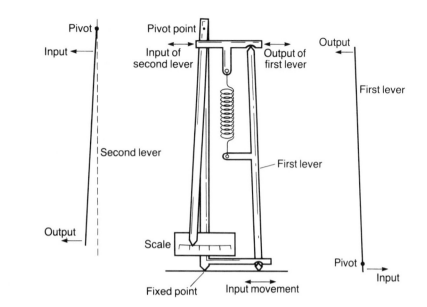

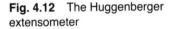

Fig. 4.12 The Huggenberger extensometer

revolution of the input shaft means that the output shaft rotates by the fraction (N_I/N_o) of a revolution. Thus the transfer function is

$$\text{transfer function} = \frac{\text{number of teeth on input gear } N_I}{\text{number of teeth on output gear } N_o}$$

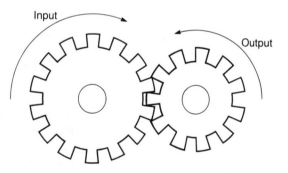

Fig. 4.13 A simple gear train signal conditioner

Compound gear trains can be used to give greater magnifications. Figure 4.14 shows the basic form of the gears used in a dial indicator gauge. The linear movement of the plunger is translated into a rotary motion by means of the rack and pinion. This causes the first gear wheel to rotate and so as a consequence cause the second gear wheel to rotate. The result of this is motion of the gear wheel on which the pointer is mounted and hence the movement of the pointer across the scale.

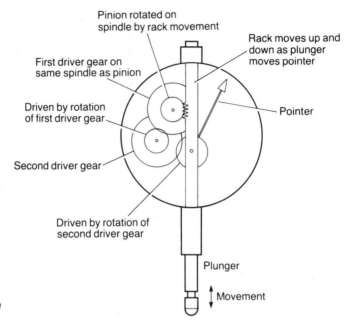

Pinion rotated on
spindle by rack movement

First driver gear on
same spindle as pinion

Rack moves up and
down as plunger
moves pointer

Driven by rotation
of first driver gear

Pointer

Second driver gear

Driven by rotation of
second driver gear

Plunger

Movement

Fig. 4.14 The dial indicator gauge

Electronic amplifier

The operational amplifier is the basic building block for both
d.c. and a.c. amplifiers. It has two inputs, known as the
inverting input (−) and the non-inverting input (+), and an
output. The output, and hence the transfer function, depends
on the connections made to these inputs. Figure 4.15(*a*) shows
the connections made to the operational amplifier when it is
used as an inverting amplifier. The input is connected to the
inverting input through a resistor R_1 and the non-inverting
input is connected to ground. A feedback path is provided
from the output terminal, via a resistor R_2, to the inverting
terminal.

The operational amplifier itself has a very large transfer
function, 100 000 or more, and the change in its output voltage
is generally limited to about ± 10 V. With this transfer function
the input voltage must be between + or − 0.0001 V. This is
virtually zero and so point X is at virtually earth potential. For
this reason it is called a virtual earth. The potential difference
across R_1 is $(V_I - V_x)$, hence the input potential V_I can be
considered to be across R_1, and so

$$V_I = I_1 R_1.$$

The operational amplifier has a very high impedance and so
virtually no current flows through X into it. Hence the current
through R_1 flows on through R_2. Because X is the virtual earth
then, since the potential difference across R_2 is $V_x - V_o$, the
potential difference across R_2 will be virtually $-V_o$. Hence

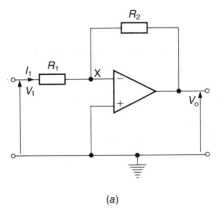

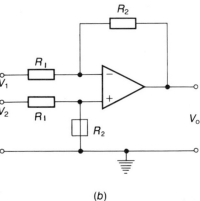

(a)

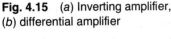

(b)

Fig. 4.15 (a) Inverting amplifier, (b) differential amplifier

$$-V_o = I_1 R_2$$

$$\text{transfer function} = \frac{V_o}{V_I} = -\frac{R_2}{R_1}$$

The transfer function is thus determined by the relative values of R_2 and R_1. The negative sign indicates that the output is inverted, i.e. 180° out of phase, with respect to the input.

Figure 4.15(b) shows the operational amplifier connected as a differential amplifier. Then

$$V_o = \frac{R_2}{R_1}(V_2 - V_1)$$

Such an amplifier finds a use in bridge circuits by amplifying the out-of-balance difference in potential (between points B and D in Figs 4.1 and 4.8).

Example 5

An inverting amplifier has a resistance of $1\,M\Omega$ in the inverting input line and a feedback resistance of $10\,M\Omega$. What is the transfer function?

Answer

Using the equation given above

$$\text{transfer function} = -\frac{R_2}{R_1} = -\frac{10}{1} = -10$$

Example 6

For the four active strain gauge bridge shown in Fig. 4.7, the gauges have a gauge factor of 2.1 and a resistance of $100\,\Omega$. When used with the load cell in the way shown, an applied load produces a strain of 1.0×10^{-5} in the compression gauges and 0.3×10^{-5} in the tensile gauges. The supply voltage for the bridge is $10\,V$. What will be the ratio of the feedback resistor resistance to that of the resistors in the two inputs of a differential amplifier used to amplify the out-of-balance potential difference if the load is to produce an output of $1\,mV$?

Answer

The change in resistance of a gauge subject to the compression is a decrease in resistance and is given by (see Ch. 3)

$$\frac{\text{change in resistance}}{\text{gauge resistance}} = -G\epsilon$$

$$\text{change in resistance} = -100 \times 2.1 \times 1.0 \times 10^{-5}$$
$$= -2.1 \times 10^{-3}\,\Omega$$

The change in resistance of a gauge subject to the tension is an increase in resistance of

change in resistance $= + 100 \times 2.1 \times 0.3 \times 10^{-5}$
$$= + 6.3 \times 10^{-4}\,\Omega$$

The out-of-balance potential difference is given by (see earlier this chapter)

$$V_{Th} = V_s\left(\frac{R_1}{R_1 + R_2} - \frac{R_3}{R_3 + R_4}\right)$$

The changes in resistance will have little effect on the denominators of the equation, being insignificant in comparison with $200\,\Omega$. Hence

$$V_{Th} = 10\left(\frac{100 + 6.3 \times 10^{-4}}{200} - \frac{100 - 2.1 \times 10^{-3}}{200}\right)$$
$$= 1.4 \times 10^{-4}\,V$$

This becomes the input to the differential amplifier, hence using the equation given earlier

$$V_o = \frac{R_2}{R_1}(V_2 - V_1)$$

with V_o to be $1\,mV$ then

$$1.0 \times 10^{-3} = \frac{R_2}{R_1}\times 1.4 \times 10^{-4}$$

and so R_2/R_1 has to be 7.1.

Signal linearization

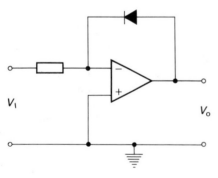

Fig. 4.16 Signal linearization circuit

Some transducers have outputs which are non-linear, e.g. the thermocouple where the thermoelectric e.m.f. is not a linear function of the temperature (this means that a graph of the e.m.f. plotted against temperature is not a straight line). One way that can often be used to turn a non-linear output into a linear one involves an operational amplifier circuit. The circuit is designed to have a non-linear relationship between its input and output so that when its input is non-linear its output is linear. This is achieved by suitable choice of components for the feedback loop.

Figure 4.16 shows such a circuit with a diode in the feedback loop. The diode has a non-linear characteristic and this results in the relationship between the input voltage V_I and the output voltage V_o becoming

$$V_o = C\ln(V_I)$$

with C being a constant. The amplifier is thus non-linear. However if the input V_I is provided by a transducer for its input θ where

$$V_I = Ke^{\alpha\theta}$$

with K and α being constants. Then

$$V_o = C\ln(Ke^{\alpha\theta})$$
$$= C\ln(K) + C\alpha\theta$$

The result is a linear relationship between the output V_o and the input to the transducer θ.

Voltage to current converter

A common feature of many process control systems is a millivolt to milliamp converter. This is required because of the widespread use of currents in the range 4–20 mA for control signals. A characteristic required of such a converter is that it has a transfer function which does not depend on the size of the voltage and also does not depend on the size of the load across the output. Figure 4.17 shows how this can be achieved with an operational amplifier. For such an arrangement

$$I_o = \left(\frac{R_2}{R_1 R_3}\right) V_I$$

$$R_1(R_3 + R_5) = R_2 R_4$$

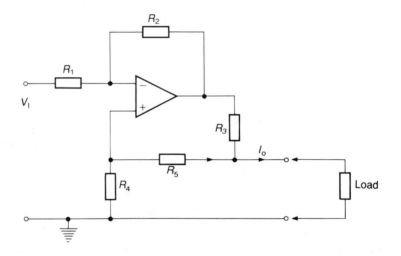

Fig. 4.17 Voltage to current converter

Current to pressure conversion

Because many control elements are pneumatic, current to pressure converters are frequently required. Figure 4.18 shows the basis of such a converter. The input current passes through coils which are then attracted towards a magnet, the extent of the attraction depending on the size of the current. This movement of the coils causes a lever to deflect. The end of the lever is a flapper for a flapper-nozzle arrangement. The position of the flapper in relation to the nozzle determines the size of the pressure signal.

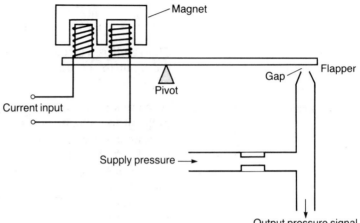

Fig. 4.18 Basis of a current to pressure converter

Attenuation

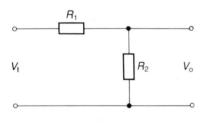

Fig. 4.19 A voltage divider circuit

An attenuator is a device which gives a voltage and/or power level output which is smaller than its input. One way of achieving this is by the use of a voltage divider circuit, Fig. 4.19 shows such a circuit. The input voltage V_I is applied across the two resistors R_1 and R_2 and the output voltage V_o taken from across the resistor R_2. Thus the fraction of the input voltage which is the output is

$$\frac{V_o}{V_I} = \frac{R_2}{R_1 + R_2}$$

The circuit in Fig. 4.19 is very basic and ignores any internal resistance of the source and resistance of the load used in the output.

Example 7

What values of resistors in the Fig. 4.19 circuit will be suitable for reducing a 20 V signal to 0.5 V?

Answer

Using the equation developed above

$$\frac{V_o}{V_I} = \frac{R_2}{R_1 + R_2}$$

Hence

$$0.5(R_1 + R_2) = 20R_2$$
$$0.5R_1 = 19.5R_2$$

If R_2 is taken to be $10\,k\Omega$ then R_1 is $390\,k\Omega$.

LIVERPOOL
JOHN MOORES UNIVERSITY
AVRIL ROBARTS LRC
TEL. 0151 231 4022

Filtering

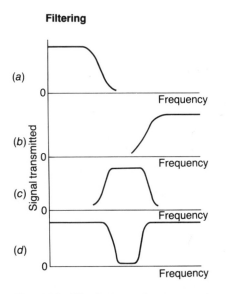

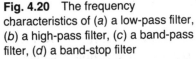

Fig. 4.20 The frequency characteristics of (a) a low-pass filter, (b) a high-pass filter, (c) a band-pass filter, (d) a band-stop filter

The term filtering is used to describe the process of removing a certain band of frequencies from a signal and permitting others to be transmitted. The range of frequencies passed by a filter is known as the *pass band*, the range not passed as the *stop band* and the boundary between stopping and passing as the *cut-off frequency*. Filters are classified according to the frequency ranges they transmit or reject. A *low-pass filter* (Fig. 4.20(a)) has a pass band in the low-frequency region, a *high-pass filter* (Fig. 4.20(b)) a pass band in the high-frequency region. A *band-pass* (Fig. 4.20(c)) filter allows a particular frequency band to be transmitted, a *band-stop* (Fig. 4.20(d)) filter stops a particular band.

The term *passive* is used to describe a filter made up using only resistors, capacitors and inductors, the term *active* is used when the filter also involves an operational amplifier. Passive filters have the disadvantage that the current that is drawn by the item that follows them can change the frequency characteristic of the filter. This problem is overcome with an active filter.

One application of filters with control and measurement systems is to improve the signal-to-noise ratio, provided the frequency spectrum of the measurement signal occupies a different frequency range from that of the noise. Thus suppose there is an interference signal at 50 Hz, a low-pass filter might be used if the signal has a lower frequency than 50 Hz or a high-pass filter if it is higher.

Modulation

A problem that is encountered with dealing with the transmission of low-level d.c. signals from transducers is that the transfer function of an operational amplifier used to amplify the d.c. signal is liable to drift. This problem can be overcome if the signal is a.c. rather than d.c. In addition, the conversion of the signal to a.c. can assist in the elimination of external interference from the transducer signal.

One way this conversion of d.c. to a.c. can be achieved is by chopping up the d.c. signal, as indicated in Fig. 4.21. The output from the chopper is a chain of pulses, the heights of which are related to the d.c. level of the input signal. This process is called *pulse amplitude modulation*. After amplification and any other signal conditioning the modulated signal might be demodulated to give a d.c. output.

With pulse amplitude modulation the height of the pulses is related to the size of the d.c. voltage. An alternative to this is *pulse width modulation*, often referred to as *pulse duration modulation*. With this the width, i.e. duration, of a pulse depends on the size of the voltage (Fig. 4.22).

LIVERPOOL
JOHN MOORES UNIVERSITY
AVRIL ROBARTS LRC
TEL. 0151 231 4022

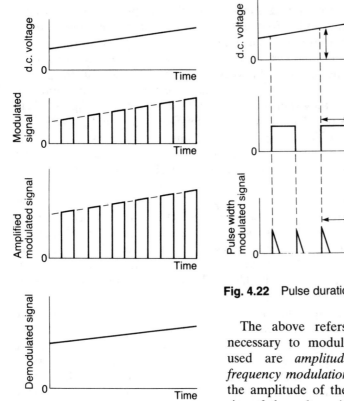

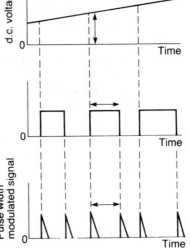

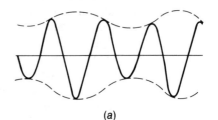

Fig. 4.22 Pulse duration modulation of a d.c. signal

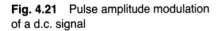

Fig. 4.21 Pulse amplitude modulation of a d.c. signal

The above refers to d.c. signals, however it is often necessary to modulate a.c. signals. Modulation techniques used are *amplitude modulation*, *phase modulation* and *frequency modulation* (Fig. 4.23). With amplitude modulation the amplitude of the carrier wave is varied according to the size of the voltage input, i.e. the wave carrying the information from the transducer. Thus for a carrier wave that can be represented by

$$v = V \sin(\omega t + \phi)$$

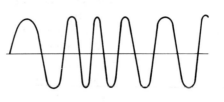

Fig. 4.23 Modulation of an a.c. transducer signal, (*a*) amplitude modulation, (*b*) frequency modulation

the amplitude term V is varied according to the way the voltage input varies. Another method of modulation involves varying the term $(\omega t + \phi)$ according to the size of the voltage input. With phase modulation the phase ϕ of the wave is varied according to the size of the voltage input. Frequency modulation involves the ω term being varied according to the size of the voltage input. Both phase modulation and frequency modulation produce similar effects, a modulated wave with a frequency which relates to the input voltage size.

Voltage to frequency conversion

When measurements are required and the transducer is some distance from the display unit, problems can occur due to the resistance of connecting leads. While a form of Wheatstone bridge can be used to compensate for lead resistance, as in Fig. 4.5, this is only suitable for relatively short lead lengths. One method of overcoming this problem is to use a voltage to frequency converter. Thus if a transducer used for the measurement of temperature is a resistance element then it could be incorporated in a Wheatstone bridge and the resulting out-of-balance potential difference converted into a frequency. This can then be transmitted to the distant display.

Analogue to digital conversion

The output from most transducers tends to be in analogue form, i.e. the size of the output from the transducer is related to the size of the input. Where a microprocessor is used as part of the measurement or control system signal processing, the analogue output from the transducer has to be converted into digital form before it can be used as an input to the microprocessor. The relationship between the input and the output for an analogue to digital conversion element can be expressed as

$$V_A \approx V_R(b_1 2^{-1} + b_2 2^{-2} + b_3 2^{-3} + \ldots b_n 2^{-n})$$

where V_A is the analogue input, V_R the reference voltage, b_1, b_2, $b_3 \ldots b_n$ the digital outputs, with n being the number of such outputs which constitute the word representing the analogue signal. Thus if, for example, the word length was limited to eight bits then n would be 8. The word length possible determines the resolution of the element, i.e. the smallest change in V_A which will result in a change in the digital output. Because of this the output from the element goes up in jumps, or steps, rather than a continuous form. It is for this reason that the equation uses the approximately equals sign $\approx$. As an illustration, consider a reference voltage of 1 V with a four-bit word. Table 4.1 shows the outcome. A change in the analogue voltage of 1/16 V means a change of 1 bit. The

Table 4.1 Analogue to digital converter table of values

V_A in V	b_1	b_2	b_3	b_4
0	0	0	0	0
1/16	0	0	0	1
2/16	0	0	0	1
3/16	0	0	1	1
4/16	0	1	0	0
5/16	0	1	0	1
etc.				
15/16	1	1	1	1

resolution is thus 1/16 V. Any change which is less than this will give no change in the digital output.

If the analogue to digital converter handles a word of length n bits then a change from 0 to 1 in b_n is the minimum change that can occur and so the resolution is

$$\text{minimum change in } V_A = V_R 2^{-n}$$

Analogue to digital converters typically have word lengths of 8, 10, 12, 14 and 16 bits.

With the 1 V reference voltage and 4-bit word the maximum value of the analogue voltage will be 15/16 V, i.e. virtually the reference voltage. The maximum value of the analogue voltage is when all the bits are ones, i.e.

$$\text{maximum } V_A = V_R(1 \times 2^{-1} + 1 \times 2^{-2} + \ldots 1 \times 2^{-n})$$

The value of the bracketed term is $(1 - 2^{-n})$. Hence

$$\text{maximum } V_A = V_R(1 - 2^{-n})$$

For a word length of 4 or more the bracketed term has a value very close to 1.

The *conversion time* of a converter is the time it requires to generate a complete digital word when supplied with the analogue input. Typically conversion times are of the order of microseconds.

See Chapter 7 and the paragraphs on digital voltmeters in the section on meters for examples of how some analogue to digital converters operate.

Example 8

A thermocouple gives an output of 0.5 mV/°C. What will be the word size and reference voltage required if the system is to be used to measure temperatures from 0–200 °C with a resolution of 0.5 °C?

Answer

At 200 °C the thermocouple output will be 200 × 0.5 = 100 mV. This would indicate a reference voltage of effectively 100 mV. The word length n required is given by

minimum change in analogue voltage = $V_R 2^{-n}$

The minimum change required is $0.5 \times 0.5 = 0.25$ mV. Hence

$0.25 = 100 \times 2^{-n}$
$\log(0.25/100) = -n \log 2$

Hence $n = 8.6$. Thus if converters with word lengths of 8, 10, 12, etc. were available a 10-bit word length one would be selected.

Digital to analogue conversion

A digital to analogue converter has an input of digital signals and an output of an analogue signal. The analogue output V_A is related to the digital input by

$$V_A = V_R(b_1 2^{-1} + b_2 2^{-2} + b_3 2^{-3} + \ldots b_n 2^{-n})$$

where V_R is the reference voltage. This is the maximum voltage the analogue output can have for the word length used. The word length is n bits, with $b_1, b_2, b_3 \ldots b_n$ being the bits. Thus for a 4-bit word length and a reference voltage of 1 V Table 4.2 shows the analogue outputs that would be obtained. The output goes up in steps of 1/16 V. The size of these steps depends on the word length and the reference voltage.

Table 4.1 Digital to analogue converter table of values

b_1	b_2	b_3	b_4	V_A in V
0	0	0	0	0
0	0	0	1	1/16
0	0	1	0	2/16
0	0	1	1	3/16
0	1	0	0	4/16
0	1	0	1	5/16
etc.				
1	1	1	1	15/16

The size of the analogue voltage increment when there is a change from 0 to 1 in b_n is

analogue voltage increment = $V_R 2^{-n}$

Thus an 8-bit converter with a 5 V reference voltage will give an analogue voltage which goes up in increments of $5 \times 2^{-8} = 0.0195$ V.

Example 9

A microprocessor gives an output of an 8-bit word. This is fed through an 8-bit digital to analogue converter to a control valve. If the control valve requires 6.0 V to be fully open what will be the reference voltage required for the converter and the percentage by which the valve opens for a change in input of 1 bit?

Answer

Using the equation given above

$$V_A = V_R(b_1 2^{-1} + b_2 2^{-2} + b_3 2^{-3} + \ldots b_n 2^{-n})$$

the maximum value of V_A will occur for the word 11111111. Thus

$$6.0 = V_R\left(\frac{1}{2} + \frac{1}{4} + \frac{1}{8} + \frac{1}{16} + \frac{1}{32} + \frac{1}{64} + \frac{1}{128} + \frac{1}{256}\right)$$

$$V_R = \frac{6.0}{0.9961} = 6.0235 \text{ V}$$

The smallest increment in analogue voltage is, using the equation given above,

$$\text{analogue voltage increment} = V_R 2^{-n}$$

$$= \frac{6.0235}{256} = 0.0235 \text{ V}$$

As a percentage this is $(0.0235/6.0235) \times 100 = 0.391\%$.

Sample and hold

A sample and hold device is needed with an analogue to digital converter to hold the analogue signal long enough for the converter to complete the conversion (Fig. 4.24). Without such a device, if the analogue signal changed during the conversion, errors can result. The device, as its name implies, takes a sample of the analogue input and holds it for the analogue to digital converter. Essentially it is a capacitor which, when switched in parallel with the input, is charged up to the analogue voltage. It then holds its potential difference until called on by the analogue to digital converter.

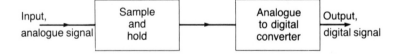

Fig. 4.24 Analogue to digital conversion

The following are some terms commonly used in the specifications of sample and hold devices. The *acquisition time* is the time taken for the capacitor to charge up to the value of the input signal. The *aperture time* is the time required for the switch to change state and switch the capacitor in or out. The *holding time* is the length of time the circuit can hold the charge without losing more than a specified percentage of its initial value. The *slew rate* is the maximum rate of input voltage that can be followed.

Multiplexers

Frequently there is a need for measurements to be sampled from a number of different locations, e.g. the temperatures in a number of rooms. Rather than use a separate measurement system for each such measurement a multiplexer can be used with a single sample-and-hold and analogue-to-digital converter (Fig. 4.25). A multiplexer is essentially a switching device which enables each of the inputs to be sampled in turn.

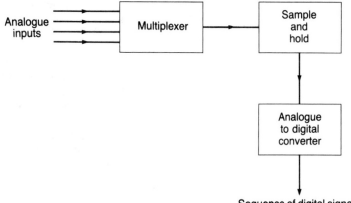

Fig. 4.25 Analogue to digital conversion using a multiplexer

Sequence of digital signals

Problems

1   A d.c. Wheatstone bridge has resistances of $20\,\Omega$ in arm BC, $500\,\Omega$ in arm CD and $200\,\Omega$ in arm AD. What will be the resistance in arm AB if the bridge is balanced?

2   A d.c. Wheatstone bridge has a 6.0 V supply connected between points A and C. What will be the potential difference between points B and D when the resistances in the bridge arms are AB $10\,\Omega$, BC $20\,\Omega$, CD $60\,\Omega$ and AD $31\,\Omega$?

3   A d.c. Wheatstone bridge has a 5.0 V supply connected between points A and C and a galvanometer of resistance $50\,\Omega$ between points B and D. What will be the current through the galvanometer when the resistances in the bridge arms are AB $120\,\Omega$, BC $120\,\Omega$, CD $120\,\Omega$ and DA $120.1\,\Omega$?

4   Explain how the Wheatstone bridge can be used to determine the resistance of the coil of a platinum resistance thermometer and compensate for any resistance changes in the leads to the coil.

5   A platinum resistance thermometer has a resistance at 0 °C of $120\,\Omega$ and forms one arm of a Wheatstone bridge. At this temperature the bridge is balanced with each of the other arms also being $120\,\Omega$. The temperature coefficient of resistance of the platinum is $0.0039\,°C^{-1}$. What will be the output voltage for change in temperature of 20 °C if the instrument used to measure it can be assumed to have infinite resistance and the supply voltage, with negligible internal resistance, for the bridge is 6.0 V?

6   For the four active strain gauge bridges shown in Fig. 4.7, the gauges have a gauge factor of 2.1 and a resistance of $120\,\Omega$. When attached to the diaphragm of a pressure gauge an applied

pressure difference from one side of the diaphragm to the other produces a strain of $+ 1.0 \times 10^{-5}$ in two of the gauges and $- 1.0 \times 10^{-5}$ in the other two. The supply voltage for the bridge is 10 V. What will be the out-of-balance potential difference output from the bridge assuming the load across the output has effectively infinite resistance?

7   Design a circuit which could be used for the cold junction compensation of a chromel–constantan thermocouple (see Ch. 3 or use tables for any data you require).

8   An a.c. bridge has in arm AB a $1.2 \text{k}\Omega$ resistor, in arm BC a $2.0 \text{k}\Omega$ resistor in series with a $1\mu\text{F}$ capacitor, in arm AD an unknown capacitance and resistance and in arm DA a $0.5 \text{k}\Omega$ resistance. What are the values of the unknown capacitance and resistance if the bridge is balanced?

9   A De Souty a.c. bridge is to be used with the capacitive liquid level gauge described in Fig. 3.12. The ratio b/a of the diameters of the concentric cylinders is 2.0 and the length of the cylinders 3.0 m. If the liquid has a relative permittivity of 2.1 what will be the value of the capacitor required in the bridge to give balance when the liquid level is 1.0 m? The resistance in the arm of the bridge opposite the gauge is 100Ω and that opposite the capacitor liquid level gauge $10 \text{k}\Omega$.

10   An inverting amplifier has a resistance of $1.2 \text{k}\Omega$ in the inverting input line and a feedback resistance of $40 \text{k}\Omega$. What is the transfer function?

11   What should be the ratio of the resistances in the input line and the feedback line for an inverting amplifier if it is required to have a transfer function of 20?

12   Explain how a flapper-nozzle arrangement can be used to provide a current to pressure converter.

13   How many bits must a digital to analogue converter have to provide outputs in increments of 0.01 V with a 5 V reference voltage?

14   Explain why an analogue to digital converter is often preceded by a sample and hold element.

15   Suggest, in the form of linked blocks, measurement systems to provide:
(a)   a digital readout of the load in a tank,
(b)   a digital readout of the temperature at a number of points in a building.

5 Controllers

Introduction

The comparison element in a control circuit has as its input the reference signal which specifies what condition is required and the feedback signal which is a measure of what condition is actually occurring. The output is the error signal. This generally acts as the input to a control unit. Its output is a signal which enables appropriate action to be taken by the correction unit to remedy the deficiency between the required and actual conditions. This might be by what is called a final control element, such as a valve. The control unit has the job of deciding what action to initiate when an error signal is received.

Commercially supplied controllers (Fig. 5.1) typically have a scale marked 0 to 100 against which the set point and feedback

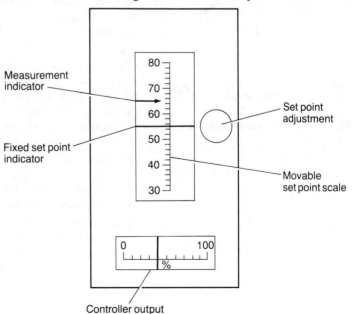

Fig. 5.1 Typical controller

signals are indicated. The scale represents the set point and measured values as percentages of the full-scale range over which the process variable can change. Adjustment of the set point is by means of a knob. The controller output is shown in a scale marked 0 to 100. The scale represents the output as a percentage of the maximum output possible.

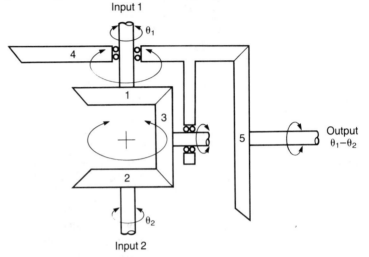

Fig. 5.2 Differential gear

Combining data

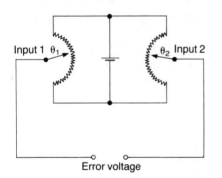

Error voltage

Fig. 5.3 Potentiometer error detector

Data can be combined in a variety of ways. Figure 5.2 shows a mechanical method, a *differential gear*, which can be used to detect angular positional errors. It consists of a number of bevel gears. If input 1 is the reference signal and fixed then gear wheel 1 is stationary. If the shaft which is input 2 (the feedback signal) rotates, the result is a motion of gear wheel 2 and consequently the intermediate bevel gear 3. This causes not only its shaft to rotate but also the gear wheel to move round the input bevel gears 1 and 2 (in Fig. 5.2 this would be out of the plane of the paper). This causes bevel gear 4 to rotate and consequently gear wheel 5. The result is an output which is proportional to the difference between the angular positions of the input shafts.

Another way of combining data about angular positions involves two *potentiometers* (Fig. 5.3). When both the potentiometers are set at the same angular positions then the output from each is the same voltage and since their outputs oppose each other the result is a zero error voltage. If the angular positions of the potentiometers differ then there is an error output voltage.

The *differential amplifier* (Fig. 4.15) or the *inverting summer* (Fig. 5.4) versions of the operational amplifier can be used to provide a combination of two data inputs. With the inverting summer

$$V_{\text{out}} = -R_{\text{f}}\left(\frac{V_1}{R_1} + \frac{V_2}{R_2}\right)$$

When $R_1 = R_2$

$$V_{\text{o}} = -\left(\frac{R_{\text{f}}}{R_1}\right)(V_1 + V_2)$$

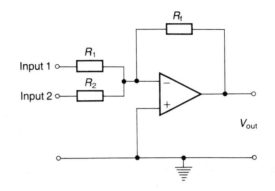

Fig. 5.4 Inverting summer

Lag

In any control system there are lags. Thus, for example, a change in the condition being controlled does not immediately produce a correcting response from the control system. This is because time is required for the system to make the necessary responses. For example, in the control of the temperature in a room by means of a central-heating system, a lag will occur between the room temperature falling below the required temperature and the control system responding and switching on the heater.

This is not the only lag. Even when the control system has responded there is a lag in the room temperature responding due to the time taken for the heat to transfer from the heater to the air in the room.

Two-step control

There are a number of ways by which a control unit can react to an error signal. The simplest way is a *two-step mode*. An example of this is the bimetallic thermostat (Fig. 5.5) used with a domestic central-heating system. If the room temperature is above the required temperature then the bimetallic strip is in an off position and the heater is off. If the room temperature falls below the required temperature then the bimetallic strip moves into an on position and the heater is switched fully on. The controller in this case can be in only two positions, on or off, as indicated by Fig. 5.6.

With the two-step mode the control action is discontinuous. A consequence of this is that oscillations occur about the required condition. This is because of lags in the time the

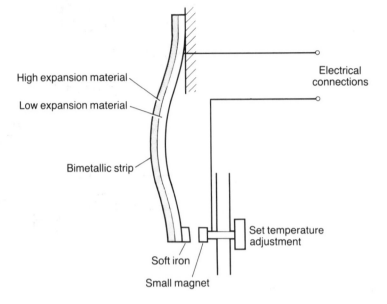

High expansion material

Low expansion material

Bimetallic strip

Electrical connections

Set temperature adjustment

Soft iron

Small magnet

Fig. 5.5 Bimetallic thermostat

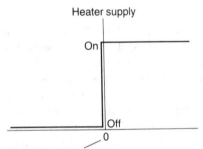

Heater supply

On

Off

0

Controller switch point

Temperature difference

Fig. 5.6 Controller characteristic

control system and the process take to respond.

For example, in the case of the domestic central-heating system when the room temperature drops below the required level the time that elapses before the control system responds and switches the heater on might be very small in comparison with the time that elapses before the heater begins to have an effect on the room temperature. In the meantime the temperature has fallen even more. The reverse situation occurs when the temperature has risen to the required temperature. Because time elapses before the control system reacts and switches the heater off, and yet more time while the heater cools and stops heating the room, the room temperature goes beyond the required value. The result is that the room temperature oscillates above and below the required temperature (Fig. 5.7).

The control is improved if instead of just a single temperature value at which the controller switches the heater on or off two values are used and the heater is switched on at a lower temperature than it is switched off (Figs 5.8 and 5.9). There is no limit to the number of values which can be provided for control action to occur, the term *multi-step* being used when there are two or more values. However the use of many steps does not avoid oscillation and so generally a controller with a continuous action is to be preferred if more than just a simple control action is required.

Two-step control action tends to be used where changes are taking place very slowly. Thus in the case of heating a room, the effect of the heater on the room temperature is only a slow change and also the cooling of the room is slow. The result of

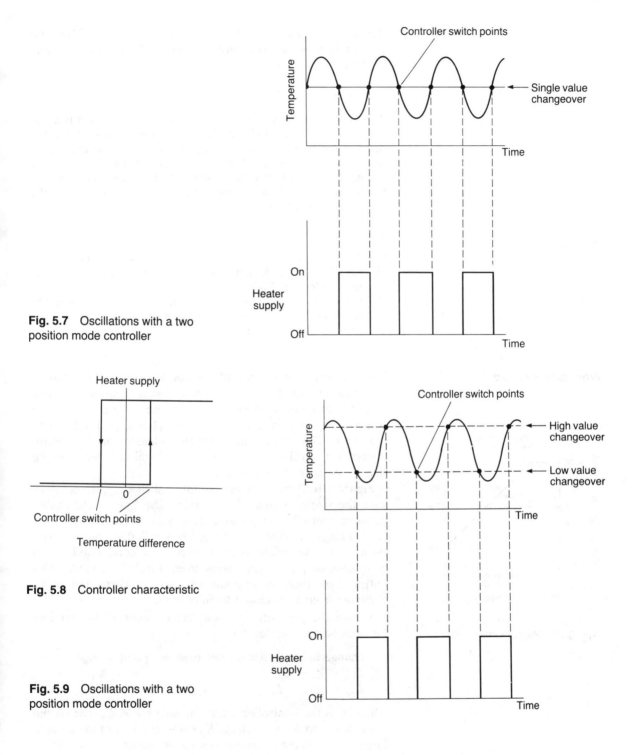

Fig. 5.7 Oscillations with a two position mode controller

Fig. 5.8 Controller characteristic

Fig. 5.9 Oscillations with a two position mode controller

this is an oscillation with a long periodic time. Two-step control is thus not very precise, but it does involve simple devices and is thus fairly cheap.

Example 1

A two-position mode controller switches on a heater when the temperature falls to 80 °C and off when it reaches 90 °C. When the heater is on the water in a container increases in temperature at the rate of 5 °C per minute, when the heater is off it cools at 2 °C per minute. If the time lags in the control system are negligible what will be the times taken from (a) the heater switching on to off, (b) the heater switching off to on?

Answer

(a) When the heater is on the temperature rises at the rate of 5 °C per minute. Thus the time taken for the temperature to rise from 80 °C to 90 °C is 2 minutes.

(b) When the heater is off the temperature falls at the rate of 2 °C per minute. Thus the time taken for the temperature to fall from 90 °C to 80 °C is 5 minutes.

Proportional control

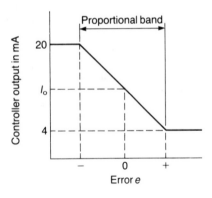

Fig. 5.10 Proportional band

With the two-step method of control, the controller output is either an on signal or an off signal, this is regardless of how great the error is. With *proportional control* the change in the controller output from the set point value is proportional to the error. This means the correction element of the control system, e.g. a valve, will receive a signal which depends on the size of the correction required.

Figure 5.10 shows how the output of such a controller varies with the size and sign of the error. The linear relationship between controller output and error tends to exist only over a certain range of errors, this range being called the *proportional band*. For the relationship shown, this corresponds to a controller output which varies from 4 mA to 20 mA. This output can then correspond to, say, a correction valve changing from fully closed to fully open.

Within the proportional band the equation of the straight line can be represented by

$$\text{change in controller output from set point} = K_p e$$
$$I_{out} - I_o = K_p e$$
$$I_{out} = K_p e + I_o$$

where I_o is the controller output at zero error, I_{out} the output at error e, and K_p a constant. K_p is the gradient of the straight line and is thus the change in controller output per unit change in error, i.e. input. Within the proportional band it is thus the transfer function of the controller. The above is the basic form of the equation that is required to describe a proportional controller.

The output of a controller is often expressed as a percentage of the full range of possible values.

Controller output as %
$$= \frac{\text{output value} - \text{minimum value}}{\text{maximum value} - \text{minimum value}} \times 100$$

Thus 0% represents the minimum controller output and 100% the maximum. Hence for Fig. 5.10, the 4 mA output is 0% and the 20 mA 100%. A 50% controller output would be 12 mA.

Similarly the error is often expressed as a percentage.

Error as %
$$= \frac{\text{measured} - \text{set point value of variable}}{\text{maximum} - \text{minimum value of variable}} \times 100$$

Thus, for example, if the measured value of the variable is say 52 °C and the required or set point temperature is 50 °C, i.e. an error of +2 °C, and the temperature range possible is 20–90 °C, the error expressed as a percentage is $+ (2/70) \times 100 = +2.9\%$. A negative value for the error would indicate a measured temperature which is below the set point temperature. A consequence of stating error as a percentage is that the proportional band is stated as a percentage.

Figure 5.11 shows how the proportional band depends on the transfer function. A high transfer function means a small proportional band. The gradient of the graph within the proportional band is K_p. Hence when the controller output and the error are expressed as percentages, the gradient is

$$\text{gradient} = K_p = \frac{100 - 0}{\text{proportional band}}$$

$$\text{proportional band} = \frac{100}{K_p}$$

Thus, for example, a requirement for a proportional band of 20% would mean that the transfer function of the controller would have to be 100/20 = 5%/%.

Generally a 50% controller output is chosen to be the output when the error is zero. Thus in the case of the controller being used to control a valve which allows water into a tank, when the error is zero the valve will be half (50%) open. This will give the normal flow rate. Any error will then increase or reduce the flow rate at a value which depends on the size of the error. The result will be to return the error to its zero value and the controller to a 50% output.

Suppose the process has the flow of liquid into a tank being

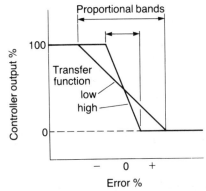

Fig. 5.11 Proportional band depends on the transfer function

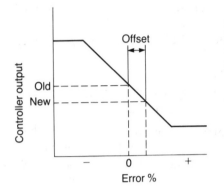

Fig. 5.12 Offset

controlled and for some reason a new set point is required for the flow rate. This could require the correcting valve to be kept open at a higher percentage, say 60%. This cannot be achieved by the zero error setting but requires a permanent error setting called an *offset* (Fig. 5.12). Such changes are said to be *load changes*.

Figure 5.13 shows the form a pneumatic proportional controller might take. When the process pressure equals the set point pressure the flapper-nozzle arrangement gives the 50% output corresponding to zero error. When the process pressure changes from this value the flapper rotates and changes the gap between flapper and nozzle. The result is a change in the output pressure. This pressure changes until the feedback bellows exerts a force to balance that due to the process bellows.

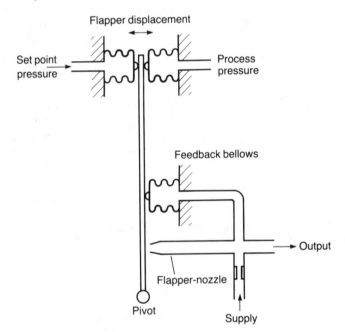

Fig. 5.13 Pneumatic proportional controller

The force due to the pressure difference between the set point and process bellows is $(P_p - P_s)A_1$, where P_p is the process pressure and P_s is the set point pressure. A_1 is the effective area of the bellows, both being assumed to be the same. The turning moment about the pivot of this force is $(P_p - P_s)A_1d_1$, where d_1 is the distance of the point of application of this force from the pivot point. The change in force due to the feedback bellows from that occurring when the set point and process pressures were equal due to the output pressure, P_{out}, is $(P_{out} - P_o)A_2$, where P_o is the value of the output pressure when there is no error and A_2 is the effective area of the feedback bellows. The turning moment about the pivot of this force is $(P_{out} - P_o)A_2d_2$ where d_2 is the distance of its point of application from the pivot point. Equilibrium occurs and the flapper stops moving when

$$(P_{out} - P_o)A_2d_2 = (P_p - P_s)A_1D_1$$

$$P_{out} = K_p(P_p - P_s) + P_o$$

where K_p is the proportionality constant and equals A_1d_1/A_2d_2. The result is the basic equation for a proportional controller.

A summing operational amplifier with an inverter can be used as a proportionality control (Fig. 5.14). For a summing amplifier we have, as indicated earlier in this chapter for Fig. 5.4

$$V_{out} = -R_f\left(\frac{V_1}{R_1} + \frac{V_2}{R_2}\right)$$

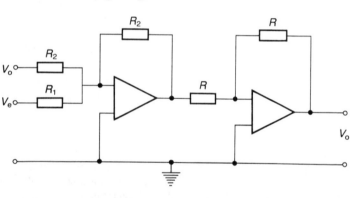

Fig. 5.14 Electronic proportionality controller

Summing amplifier   Inverter

But when the feedback resistor $R_f = R_2$, input 1 is the zero error voltage value V_o, and input 2 is the error signal V_e, as in Fig. 5.14, then the equation becomes

$$V_{out} = -R_2\left(\frac{V_o}{R_2} + \frac{V_e}{R_1}\right)$$

$$V_{out} = -\frac{R_2}{R_1}V_e - V_o$$

If this output from the summing amplifier is then passed through an inverter, i.e. an operational amplifier with a feedback resistance equal to the input resistance and hence a transfer function of -1, then

$$V_{out} = \frac{R_2}{R_1}V_e + V_o$$

$$V_{out} = K_pV_e + V_o$$

where K_p is the proportionality constant. The result is the basic equation for a proportional controller.

Example 2

A proportional controller is used to control the height of water in a tank where the water level can vary from 0–9.0 m. What proportional band and transfer function will be required if the required height of water is 5.0 m and the controller is to fully close a valve when the water rises to 5.5 m and fully open it when the water falls to 4.5 m?

Answer

When the error is -0.5 m the controller output must be 100% open and when $+0.5$ m 0% open. The proportional band must therefore extend from a height error of -0.5 m to $+0.5$ m. Expressed as a percentage, the proportional band extends from $-(0.5/9.0) \times 100 = -5.6\%$ to $+(0.5/9.0) \times 100 = +5.6\%$. The proportional band is thus 11.2%.

This value of proportional band will mean a transfer function K_p which is given by

$$\text{proportional band} = \frac{100}{K_p}$$

$$K_p = 100/11.2 = 8.9$$

Note that the units of the transfer function are % per %.

Example 3

A proportional controller has a transfer function of 15%/% and a set point of 50% output. It outputs to a valve which at the set point allows a flow of 200 m³/s. The valve changes its output by 4 m³/s for each percent change in controller output. What will be the controller output and the offset error when the flow has to be changed to 240 m³/s?

Answer

The new controller setting for a flow of 240 m³/s must be 240/4 = 60%.

When the controller output and the error are expressed as percentages the gradient of the proportionality line is

$$K_p = \frac{\text{change in controller output}}{\text{change in error}}$$

with K_p being expressed as %/%. Hence

$$15 = \frac{60 - 50}{e}$$

$$e = 0.67\%$$

This is the offset error.

Derivative control

With *derivative control* the change in controller output from the set point value is proportional to the rate of change with time of the error signal. This can be represented by the equation

$$P_{\text{out}} - P_{\text{o}} = K_{\text{D}} \frac{\text{d}e}{\text{d}t}$$

where P_{o} is the set point output value, P_{out} the output value that will occur when the error e is changing at the rate $\text{d}e/\text{d}t$. K_{D} is the constant of proportionality, sometimes referred to as the derivative transfer function or more commonly the derivative time since it has units of time. This equation can be rearranged to

$$P_{\text{out}} = K_{\text{D}} \frac{\text{d}e}{\text{d}t} + P_{\text{o}}$$

It is usual to express the controller output as a percentage of the maximum minus minimum output and the error as a percentage of the full range of the variable. Thus K_{D} is in the units % second per %. Figure 5.15 shows the graph of this equation.

With derivative control as soon as the error signal begins to change there can be quite a large controller output since it is proportional to the rate of change of the error signal and not its value. Rapid initial response to error signals thus occurs. Figure 5.16 shows the controller output that results when there is a constant rate of change of error signal with time. The controller output is constant because the rate of change is constant and occurs immediately the deviation occurs.

Derivative controllers do not however respond to steady state error signals, since with a steady error the rate of change of error with time is zero. Because of this derivative control is often combined with proportional control. Figure 5.17 shows the form of an electronic derivative controller circuit. K_{D} is R_2C.

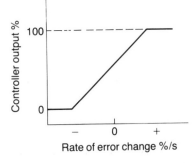

Fig. 5.15 Derivative mode of control

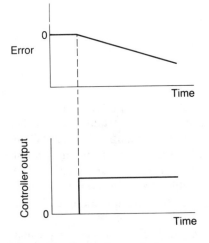

Fig. 5.16 Derivative control

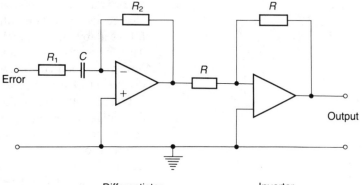

Fig. 5.17 Electronic derivative control

Example 4

A derivative controller has a set point of 50% and derivative constant K_D of 0.4% s/%. What will be the controller output when the error (*a*) changes at 1%/s, (*b*) is constant at 4%?

Answer

(*a*) Using the equation given above

$$P_{out} = K_D \frac{de}{dt} + P_o$$

$$= 0.4 \times 1 + 50 = 50.4\%$$

(*b*) With de/dt zero then P_{out} equals P_o, i.e. 50%. The output only differs from the set point value when the error is changing.

Proportional plus derivative control

With proportional plus derivative control the change in controller output from the set point value is given by

$$\text{change in controller output from set point} = K_p\left(e + K_D\frac{de}{dt}\right)$$

$$P_{out} - P_o = K_p\left(e + K_D\frac{de}{dt}\right)$$

$$P_{out} = K_p\left(e + K_D\frac{de}{dt}\right) + P_o$$

where P_o is the output at the set point, P_{out} the output when the error is e, K_p is the proportionality constant and K_D the derivative constant, de/dt is the rate of change of error. When the outputs are expressed as percentages of the maximum minus minimum output, and the error as a percentage of the full range of values of the variable, then the above equation can be used with percentages and then K_p has the unit % per % and K_D % second per %.

Figure 5.18 shows how the controller output can vary when there is a constantly changing error. There is an initial quick change in controller output because of the derivative action followed by the gradual change due to proportional action. This form of control can thus deal with fast process changes, however a load change will require an offset error (see earlier discussion of proportional control).

Figure 5.19 shows the basic form of a proportional plus derivative pneumatic controller. The restriction in the air supply to the proportional bellows means that it cannot respond quickly to air pressure changes. Thus when the flapper moves as a consequence of a pressure difference between the set point bellows and the process bellows the air

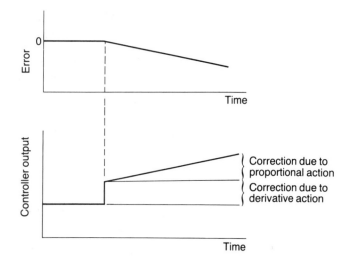

Fig. 5.18 Proportional plus derivative control

escaping from the nozzle changes. The consequential change in the output pressure is a rather rapid change since the restriction prevents the proportional bellows quickly responding. The result is a change which is proportional to the rate of change of the pressure difference between the set point and process bellows. With time the proportional bellows responds and a further change takes place which is proportional to the pressure difference between the set point and process bellows. The restriction in the line to the proportional bellows is generally variable since it determines the value of K_D.

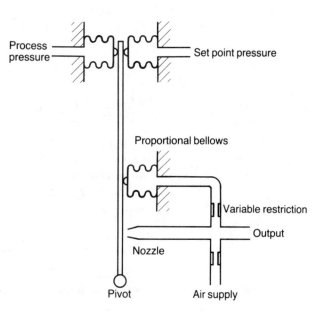

Fig. 5.19 Proportional plus derivative pneumatic controller

Example 5

What is the controller output for a proportional plus derivative controller initially and 2 s after the error begins to change from the zero error at the rate of 1.2%/s? The controller has a set point of 50%, $K_p = 4$ and $K_D = 0.4\%$ s/%.

Answer

Using the equation given above

$$P_{out} = K_p\left(e + K_D\frac{de}{dt}\right) + P_o$$

Initially the error e is zero. Hence

$$P_{out} = 4(0 + 0.4 \times 1.2) + 50 = 51.9\%$$

Because the rate of change is constant, after 2 s the error will have become 2.4%. Hence

$$P_{out} = 4(2.4 + 0.4 \times 1.2) + 50 = 54.6\%$$

Integral control

Integral control is where the rate of change of the control output P is proportional to the input error signal e.

$$\frac{dP}{dt} = K_I e$$

K_I is the constant of proportionality and, when the controller output is expressed as a percentage of the maximum minus minimum output and the error as a percentage of the range of the variable quantity, has units of % per %s. Integrating the equation gives

$$\int_{P_o}^{P_{out}} dP = \int_0^t K_I e \, dt$$

$$P_{out} - P_o = \int_0^t K_I e \, dt$$

$$P_{out} = \int_0^t K_I e \, dt + P_o$$

P_o is the controller output at zero time, P_{out} is the output at time t. The equation can be used with all quantities expressed as percentages.

The reciprocal of K_I is called the *integral time* T_I and is in seconds.

$$\text{Integral time } T_I = \frac{1}{K_I}$$

Figure 5.20 illustrates the action of an integral controller. We can consider the graphs in two ways. When the controller

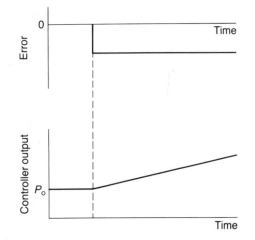

Fig. 5.20 Integral control

output is constant, i.e. dP/dt is zero, then the error is zero. When the controller output varies at a constant rate, i.e. dP/dt is constant, then the error must have a constant value. The alternative way of considering the graphs is in terms of the area under the error graph.

$$\int_0^t e\,dt = \text{area under the error graph between zero time and } t$$

Thus up to the time when the error occurs the value of the integral is zero. Hence $P_{out} = P_o$. When the error occurs it maintains a constant value. Thus the area under the graph is increasing as the time increases. Since the area increases at a constant rate the controller output increases at a constant rate.

The integral mode of control is not usually used alone but is frequently used in conjunction with the proportional mode. Figure 5.21 shows the form of the circuit used for an electronic integral controller. K_I is $1/R_1C$.

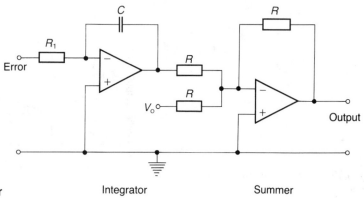

Fig. 5.21 Electronic integral controller

Example 6

An integral controller, with a value of K_I of 0.10%/% s, has an output of 40% at the set point. What will be the output after (a) 1 s, (b) 2 s, if there is a sudden change to a constant error of 20%?

Answer

Using the equation developed above

$$P_{out} = \int_0^t K_I e \, dt + P_o$$

When the error does not vary with time the equation becomes

$$P_{out} = K_I e t + P_o$$

Thus for (a) when $t = 1$ s,

$$P_{out} = 0.10 \times 20 \times 1 + 40 = 42\%$$

For (b) when $t = 2$ s,

$$P_{out} = 0.10 \times 20 \times 2 + 40 = 44\%$$

Proportional plus integral control

When integral action is added to a proportional control system the controller output P_{out} is given by

$$P_{out} = K_p(e + K_I \int e \, dt) + P_o$$

where K_p is the proportional control constant, K_I the integral control constant, P_{out} the output when there is an error e and P_o the output at the set point when the error is zero. Figure 5.22 shows how the system reacts when there is an abrupt change to a constant error. The error gives rise to a proportional controller output which remains constant since

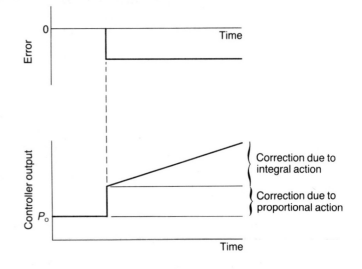

Fig. 5.22 Integral plus proportional action

the error does not change. There is then superimposed on this a steadily increasing controller output due to the integral action.

Suppose there is a requirement to change the controller set point from, say, 50% to 60% to allow for some load change in the process. With just a proportional controller this can only be done by having an offset error, i.e. an error value other than zero for the set point value. However with the combination of integral and proportional control this is not the case. The integral part of the control can provide a change in controller output without an offset error. The controller can be said to reset its set point. Figure 5.23 shows the effects of the proportional action and the integral action if we create an error signal which is increased from the zero value and then decreased back to it again. With proportional action alone the result is the controller mirrors the change and ends up back at its original set point value. With the integral action the controller output increases in proportion to the way the area under the error–time graph increases and since, even when the error has reverted back to zero, there is still a value for the area there is a change in controller output which continues after the error has ceased. The result of combining the proportional and integral actions, i.e. adding the two separate graphs, is thus a change in controller output without an offset error.

Because of the lack of an offset error, this type of controller can be used where there are large changes in the process variable. However because the integration part of the control takes time the changes must be relatively slow to prevent oscillations. Another disadvantage of this form of control is that when the process is started up with controller output at 100%, e.g. with a liquid level control the initial condition may be an empty tank and so the error is so large that the controller has to give a 100% output to fully open a valve, the integral action causes a considerable overshoot of the error before finally settling down.

Figure 5.24 shows the basic form of a proportional plus integral pneumatic controller. A difference in pressure between the process and set point bellows results in a movement of the flapper. This changes the output pressure and the pressure in the proportional bellows. The result is a motion of the flapper until the turning moment resulting from the force exerted by the proportional bellows balances that resulting from the difference between the process and set point bellows. While this is happening the integral bellows has barely been affected because of the time delay introduced by the restriction. However as the integral bellows slowly comes up to the output pressure it moves the flapper and so changes the output pressure.

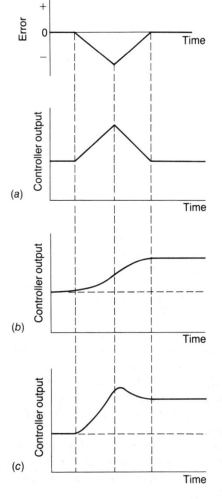

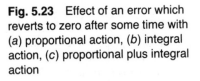

Fig. 5.23 Effect of an error which reverts to zero after some time with (a) proportional action, (b) integral action, (c) proportional plus integral action

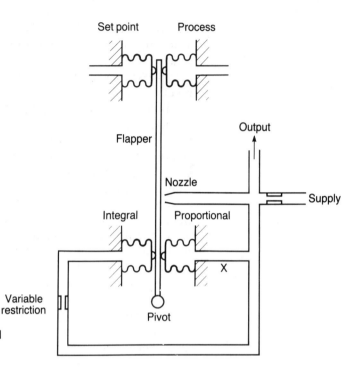

Fig. 5.24 Proportional–integral pneumatic controller

Proportional plus integral plus derivative control

Combining all three modes of control enables a controller to be produced which has no offset error and reduces the tendency for oscillations. Such a controller is known as a *three mode controller*. The equation describing its action is

$$P_{out} = K_p\left(e + K_I \int e \, dt + K_D \frac{de}{dt}\right) + P_o$$

where P_{out} is the output from the controller when there is an error e which is changing with time t, P_o is the set point output when there is no error, K_p is the proportionality constant, K_I the integral constant and K_D the derivative constant. One way of considering a three mode controller is as a proportional controller which has integral control to eliminate the offset error and derivative control to reduce time lags.

Figure 5.24 with another variable constriction at point X will give a three-mode pneumatic controller. Figure 5.25 shows the basic form of a three-mode electronic controller. K_p is R_2/R_1, K_I is $1/R_I C_I$, and K_D is $R_D C_D$.

Example 7

What will be the controller output of a three-mode controller having K_p as 4%/%, K_I as 0.6%/% s, K_D as 0.5% s/%, a set point output of 50% and subject to the error change shown in Fig. 5.26, (*a*) immediately the change starts to occur and (*b*) 2 s after it starts?

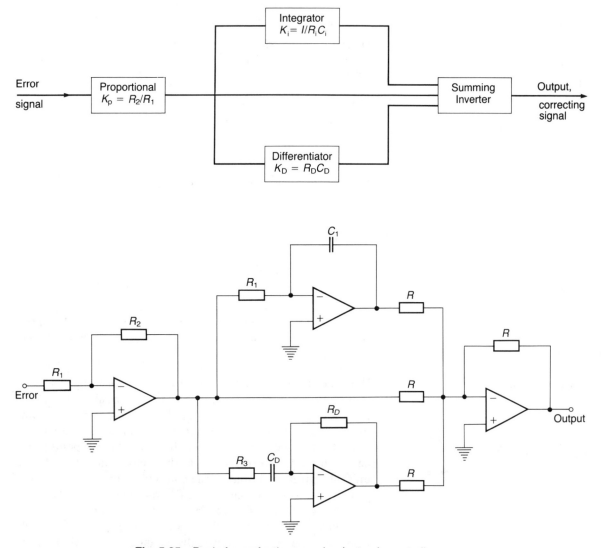

Fig. 5.25 Basic form of a three mode electronic controller

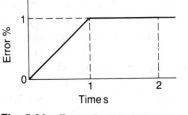

Fig. 5.26 Example 7

Answer

Using the equation given above

$$P_{out} = K_p\left(e + K_I\int e \,dt + K_D\frac{de}{dt}\right) + P_o$$

(a) Initially $e = 0$, $de/dt = 1\%/s$, $\int e \,dt = 0$

Thus $P_{out} = 4(0 + 0 + 0.5 \times 1) + 50 = 52.0\%$

(b) At 2 s, $e = 1\%$, $\int e \,dt = 1.5\%$ s, $de/dt = 0$

The integral is the area under the error–time graph up to 2 s.
Thus $P_{out} = 4(1 + 0.6 \times 1.5 + 0) + 50 = 57.6\%$

Example 8

Design a three mode controller using operational amplifiers so that K_P is 2, K_I 0.1/s, and K_D 30 s. Input and output are equally scaled voltages.

Answer

Using the circuit shown in Fig. 5.25,

$$K_p = 2 = \frac{R_2}{R_1}$$

Hence if R_1 is chosen to be $1\,k\Omega$ then R_2 has to be $2\,k\Omega$.

$$K_I = 0.1 = \frac{1}{R_I C_I}$$

Thus if C_I is chosen to be $20\,\mu F$ then R_I has to be $500\,k\Omega$.

$$K_D = 30 = R_D C_D$$

Thus if C_D is chosen to be $20\,\mu F$ then R_D must be $1.5\,M\Omega$.

The statement about the input and output being equally scaled voltages means that the voltage range of the output equals the voltage range of the input. The circuit as given in Fig. 5.25 is based on this.

Cascade control

With the single feedback loop form of control so far discussed in this chapter a single variable, say level of liquid in a tank, of a process is controlled (Fig. 5.27) by a transducer responding to the level, feeding back a signal to an error detector which finds the difference between this signal and the set point signal and then sends an error signal to the control unit. The control unit then gives an output to the correcting unit and this in turn produces a correcting change in the variable, i.e. the level. In controlling the liquid level in a tank by means of a single transducer and single feedback loop, the controller can only respond when the level actually changes. Only then, after some time lag, will the controller change the opening of the valve allowing liquid into the tank. The level might change because of a change of the flow-rate of liquid entering the tank but there can be no response until the level in the tank changes. The single-loop system cannot anticipate changes in level occurring.

A better control system, which reduces time lags, the effects of load changes and other disturbances is cascade control. *Cascade control* involves the use of two controllers and two feedback loops (Fig. 5.28). The outer loop or main loop is concerned with the control of the main variable, in this case the level of liquid in the tank. The inner loop or minor loop is concerned with some intermediate variable, such as the flow rate of liquid entering the tank. The set point of the outer loop

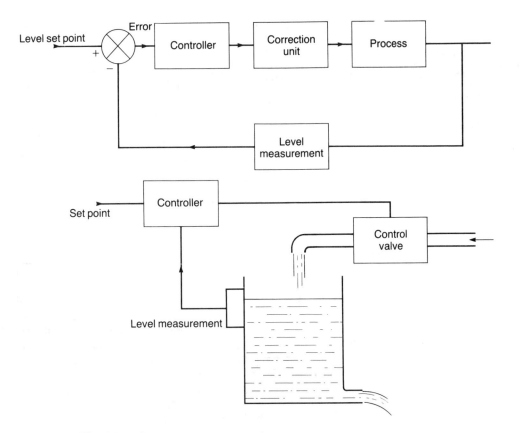

Fig. 5.27   Single loop control of liquid level in a tank

is the required level and is set by the person in charge of the process. The set point of the inner loop is however determined by the outer loop controller. This means, since the output of the outer loop controller is determined by the error signal it receives, that the set point is determined by the measurement made of the level of liquid in the tank.

With such an arrangement, if there is a change in the supply of fluid in the pipe to the tank then the flow measurement indicates this and sends a signal to the inner controller. The result is an error signal and so an output from the controller which changes the control valve opening to correct the change before the liquid leaves the pipe and enters the tank. If the liquid level changes, perhaps as a result of more being drawn from the tank, then the level measurement leads to an error signal to the outer controller which then changes the set point of the inner controller and so its output to the valve. Hence the effect of a supply change is corrected near to its source and time lags are reduced.

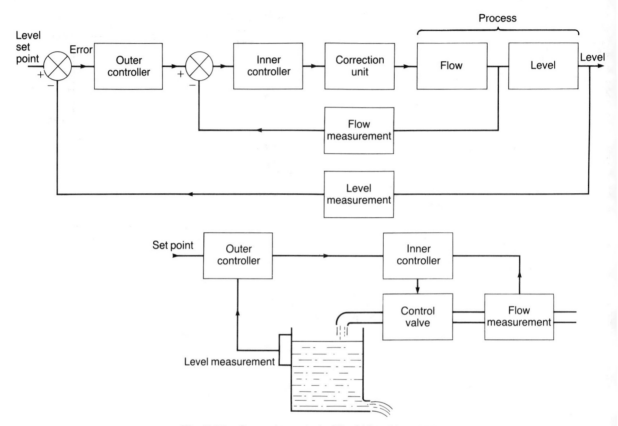

Fig. 5.28 Cascade control of liquid level in a tank

Digital control

A digital computer used as a controller requires inputs which are digital, processes the information in digital form and gives an output in digital form. Since many control systems have analogue measurements and require analogue outputs to correcting elements, analogue-to-digital converters will be required for the inputs to the computer and digital-to-analogue converters for the outputs. Digital computers as controllers have the advantage over analogue controllers that the form of the controlling action, e.g. proportional or three mode, can be altered by purely a change in the computer software. No changes in hardware or electrical wiring is required. Indeed the control strategy can be altered by the computer program during the control action in response to the developing situation.

They also have other advantages. With analogue control separate controllers are required for each process being controlled. With a digital computer many separate processes can be controlled by sampling process with a multiplexer (see Ch. 4). Digital control gives better accuracy than analogue control because the amplifiers and other components used with analogue systems change thei. characteristics with time and temperature and so show drift while digital control because it operates on signals in only the on-off mode does not suffer from drift in the same way.

Tuning

The term *tuning* is used to describe the process of selecting the best controller settings. There are a number of methods of doing this, here just two methods will be discussed, both by Ziegler and Nichols. The first method is often called the *process reaction curve method*. The procedure with this method is to open the process control loop so that no control action occurs. Generally the break is made between the controller and the correction unit. A test input signal is then applied to the correction unit and the response of the measured process variable determined, i.e. the error signal. The test signal should be as small as possible. Figure 5.29 shows the form of test signal and a typical response. The graph of measured signal plotted against time is called the *process reaction curve*.

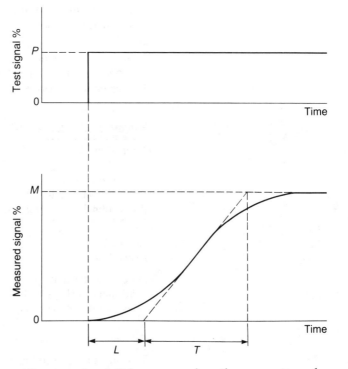

Fig. 5.29 Process reaction curve

The test signal P is expressed as the percentage change in the correction unit. The measured variable is expressed as the percentage of the full-scale range. A tangent is drawn to give the maximum gradient of the graph. For Fig. 5.29 the maximum gradient R is M/T. The time between when the test signal started and where this tangent intersects the graph time axis is termed the lag L. Table 5.1 gives the criteria recommended by Ziegler and Nichols for control settings based on the values of P, R and L.

Table 5.1 Ziegler and Nichols: process reaction curve criteria

Control mode	K_p	K_I	K_D
Proportional only	P/RL		
Proportional + integral	$0.9\,P/RL$	$1/3.33\,L$	
Proportional + integral + derivative	$1.2\,P/RL$	$1/2\,L$	$0.5\,L$

Note: proportional band = $100/K_p\%$.

The other method is called the *ultimate cycle method*. Firstly integral and derivative actions are reduced to their minimum values. The proportional constant K_p is set low and then gradually increased. This is the same as saying that the proportional band is gradually made narrower. While doing this small disturbances are applied to the system. This is continued until continuous oscillations occur. The critical value of the proportional constant K_{pc} at which this occurs is noted and the periodic time of the oscillations T_c measured. Table 5.2 shows how the Ziegler and Nichols recommended criteria for controller settings are related to this value of K_{pc}. The critical proportional band is $100/K_{pc}$.

Table 5.2 Ziegler and Nichols: ultimate cycle criteria

Control mode	K_p	K_I	K_D
Proportional only	$0.5\,K_{pc}$		
Proportional + integral	$0.45\,K_{pc}$	$1.2/T_c$	
Proportional + integral + derivative	$0.6\,K_{pc}$	$2.0/T_c$	$T_c/8$

Example 9

Determine the settings of K_p, K_I and K_D required for a three mode controller which gave the process reaction curve shown in Fig. 5.30 when the test signal was a 6% change in the control valve position.

Answer

Drawing a tangent to the maximum gradient part of the graph gives a lag L of 150s and a gradient R of $5/300 = 0.017\%/s$. Hence

$$K_P = \frac{1.2P}{RL} = \frac{1.2 \times 6}{0.017 \times 150} = 2.82\%/\%$$

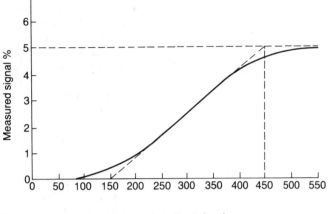

Fig. 5.30 Example 9

Time from start of test signal s

$$K_{I} = \frac{1}{1.2\,L} = \frac{1}{1.2 \times 150} = 0.0056\ \text{s}$$

$$K_{D} = 0.5\,L = 0.5 \times 150 = 75\ \text{s}$$

Example 10

When tuning a three-mode control system by the ultimate cycle method it was found that oscillations begin when the proportional band is decreased to 30%. The oscillations have a periodic time of 500 s. What are the suitable values of K_p, K_I and K_D?

Answer

The critical value of K_{pc} is 100/critical proportional band and so $100/30 = 3.33$.

Using the criteria given in Table 5.2,

$$K_p = 0.6\,K_{pc} = 0.6 \times 3.33 = 2.0\%/\%$$

$$K_I = \frac{2}{T_c} = \frac{2}{500} = 0.004\%/\%\ \text{s}$$

$$K_D = \frac{T_c}{8} = \frac{500}{8} = 62.5\ \text{s}$$

Problems

1 Explain the purpose of a control system comparing the set point value of some variable with the measured value and describe how this comparison may be made with electrical signals.

2 What are the limitations of two-step (on-off) control and in what situations is such a control system commonly used?

3 A two-position mode controller switches on a room heater when the temperature falls to 20 °C and off when it reaches 24 °C. When the heater is on the air in the room increases in temperature at the rate of 0.5 °C per minute, when the heater is off it cools at 0.2 °C per minute. If the time lags in the control system are negligible what will be the times taken from (*a*) the heater switching on to off, (*b*) the heater switching off to on?

4 A two-position mode controller is used to control the water level in a tank by opening or closing a valve which in the open position allows water at the rate of $0.4\,m^3/s$ to enter the tank. The tank has a cross-sectional area of $12\,m^2$ and water leaves it at the constant rate of $0.2\,m^3/s$. The valve opens when the water level reaches 4.0 m and closes at 4.4 m. What will be the time taken from (a) the valve opening to closing, (b) the valve closing to opening?

5 Explain what is meant by proportional action in a process controller and the term offset.

6 A proportional controller is used to control the height of water in a tank where the water level can vary from zero to 4.0 m. The required height of water is 3.5 m and the the controller is to fully close a valve when the water rises to 3.9 m and fully open it when the water falls to 3.1 m. What proportional band and transfer function will be required?

7 A proportional controller has K_p of 20%/% and a set point of 50% output. Its output is to a valve which at the set point allows a flow of $2.0\,m^3/s$. The valve changes its output in direct proportion to the controller output. What will be the controller output and the offset error when the flow has to be changed to $2.5\,m^3/s$?

8 A derivative controller has a set point of 50% and derivative constant K_D of 0.5% s/%. The error starts at zero and then changes at 2%/s for 3 s before becoming constant for 2 s, after which it decreases at 1%/s to zero. What will be the controller output at (a) 0 s, (b) 1 s, (c) 4 s, (d) 6 s?

9 An integral controller has a set point of 50% and K_L of 0.10%/% s. The error starts at zero and changes at 4%/s for 2 s before becoming constant for 3 s. What will be the output after (a) 1 s, (b) 3 s?

10 A three-mode controller has K_p as 2%/%, K_I as 0.1%/% s, K_D as 1.0% s/%, and a set point output of 50%. The error starts at zero and changes at 5%/s for 2 s before becoming constant for 3 s. It then decreases at 2%/s to zero and remains at zero. What will be the controller output at (a) 0 s, (b) 3 s, (c) 7 s, (d) 11 s?

11 Describe and compare the characteristics of (a) proportional control, (b) proportional plus integral control, (c) proportional plus integral plus derivative control.

12 Design a three-mode controller using operational amplifiers and which will have a 40% proportional band, K_I 0.08 s and K_D 50 s. Input and output are equally scaled voltages.

13 Determine the settings of K_p, K_I and K_D required for a three mode controller which gave a process reaction curve with a lag L of 200 s and a gradient R of 0.010%/s when the test signal was a 5% change in the control valve position.

14 When tuning a three-mode control system by the ultimate cycle method it was found that oscillations begin when the proportional band is decreased to 20% with the oscillations having a periodic time of 200 s. What are the suitable values of K_p, K_I and K_D?

15 Describe a procedure that could be used to tune a control system.

6 Correction units

The final control operation

The correction unit has as its input a signal from the controller. This signal carries the necessary information to enable corrective action to be taken to reduce the error between the measured and the set value of the process variable. The output from the correction unit is some change in the process. Thus, for example, a correction unit may have an input from the control unit of a signal between 4 and 20 mA and give an output of a change in a flow rate of some liquid.

In general, correction units are often made up of three elements.

1 *Signal conversion* The signal from the controller may need some modification to make it suitable for the next element. For example, there may need to be a current to voltage conversion or perhaps signal amplification.

2 *Actuators* The actuator is a device which translates the converted signal into action on the control element. For example, if the control element is a valve the actuator may be a device which converts a voltage signal into an opening or closing of the valve. Another example is a relay which converts a current signal into the switching on or off of the control element, possibly an electric heater.

3 *Control element* This is the device which has a direct influence on the process variable. It might be an electric heater which changes the process variable of temperature or a valve to change the flow rate of a liquid into a tank and hence the process variable of liquid level.

Signal conversions

A typical controller output is a current in the range 4–20 mA. This has often to be converted into a more suitable form for an actuator. Thus if the actuator is a motor the current may need to be converted into a voltage. In addition there is generally a power conversion since the output of a controller is a low-power signal and actuators generally require high-power signals. Thus with the motor there will be not only a current to voltage conversion but also power amplification.

Typical signal convertors (see Ch. 4 for details) are:

1. Electronic d.c. and a.c. amplifiers.
2. Magnetic amplifiers.
2. Digital to analogue converter.
4. Current to pressure converter.

Actuators

Actuators can be classified as being electrical, pneumatic or hydraulic. Electric actuators respond to inputs of electrical signals, pneumatic and hydraulic actuators to pressure signals. Common electrical actuators are relays, solenoids, a.c. and d.c. motors, and digital stepping motors. Pneumatic and hydraulic actuators enable large forces to be produced from relatively small pressure changes. These forces are often used with valves to move diaphragms and so affect, for example, the flow of liquid through the valve.

Electric actuators

The following are commonly used electric actuators.

Relays

The electric relay offers a simple on/off action in response to a control signal (Fig. 6.1). Thus a relay might be used to switch power on or off to an electric heater in order to control the temperature in some process.

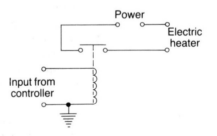

Fig. 6.1 A relay as an actuator

Solenoids

The actuator consists of a solenoid within which a soft iron plunger can move (Fig. 6.2). Changing the current through the solenoid changes the position of the plunger. This movement might be magnified by a system of levers.

Direct-current motors

These are widely used for positional or speed-control systems. The field and armature windings are separately supplied. Thus, for example, a constant voltage may be applied to the field windings to produce the magnetic field and the output voltage from the controller (probably via a signal conversion

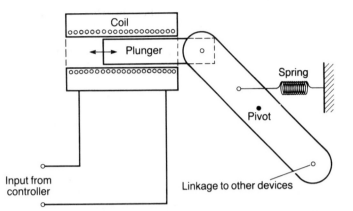

Fig. 6.2 A solenoid as an actuator

unit) applied to the armature windings (Fig. 6.3). The result is that the rotation of the armature which is proportional to the armature current is proportional to the controller output. An alternative to this form of connection is to apply the controller voltage to the field windings and a constant voltage to the armature windings.

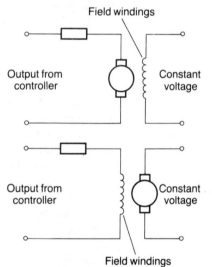

Fig. 6.3 A d.c. motor as an actuator

Stepping motors

The *stepping motor* is a device that rotates through some angle for each digital pulse supplied to its input. Thus, for example, if such a motor requires 60 pulses to rotate through 360° then 1 pulse causes a rotation of 6°. If there is an input of 120 pulses per second then it rotates through 2 revolutions per second. A stepper motor consists of a number of poles, each carrying a field winding, with a rotor which is often a permanent magnet (Fig. 6.4(*a*). By inputting to each opposite pairs of field windings in turn the axis of the magnetic field rotates and the

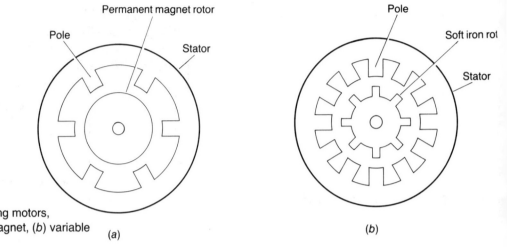

Fig. 6.4 Stepping motors, (a) permanent magnet, (b) variable reluctance

(a)

(b)

rotor axis follows it. Figure 6.4(*b*) shows another form, the *variable reluctance stepping motor*. The rotor is made of soft iron and cylindrical with slots. It always moves to a position which gives the minimum reluctance. This position depends upon which pairs of the field windings happen at that moment to be carrying currents.

Pneumatic and hydraulic actuators   The most common form of pneumatic actuator is that associated with control valves, the *diaphragm actuator*. Essentially it consists of a diaphragm with the input pressure from the controller, typically between 10–100 kPa, on one side and atmospheric pressure on the other. The diaphragm is made of rubber which is sandwiched in its centre between two circular steel dishes. The effect of changing controller pressure is thus to move the central part of the diaphragm, as illustrated in Fig. 6.5. This is communicated to the valve by a shaft which is attached to the diaphragm. A restoring force is provided by a spring.

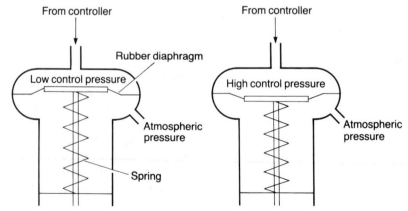

Fig. 6.5   Pneumatic diaphragm actuator

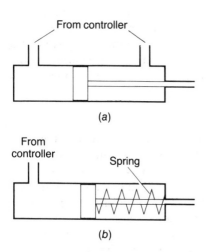

Fig. 6.6 Hydraulic cylinders, (a) double acting, (b) single acting with spring return

The pneumatic diaphragm actuator is often referred to as a *linear actuator* since the signal from the controller is converted into linear motion. The *hydraulic or pneumatic cylinder* is another example. The principles and form are the same for both hydraulic and pneumatic versions, differences being purely a matter of size as a consequence of the higher pressures used with hydraulics (up to of the order of 30 MPa compared with less than 700 kPa for pneumatics). The controller used with the hydraulic cylinder is the *spool valve*.

The hydraulic cylinder consists of a hollow cylindrical tube along which a piston can slide (Fig. 6.6). The term *double-acting* is used when pressures are applied to each side of the piston. A difference in pressures between the two sides of the piston results in motion of the piston, it being able to move in either direction along the cylinder. The term *single-acting* is used when the fluid pressure is applied to just one side of the piston. The piston in this case can move in only one direction, a spring being frequently used to give a return stroke.

Figure 6.7 shows the combination of a *spool valve* with a hydraulic cylinder, the combination often being known as a *hydraulic servomotor* or *hydraulic amplifier*. The spool valve is

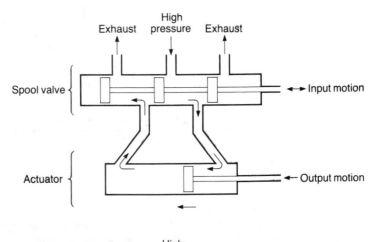

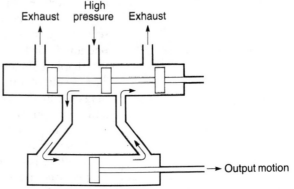

Fig. 6.7 Spool valve with hydraulic cylinder

essentially a controller. It has a constant high-pressure supply which is switched to either side of the actuator piston by the input motion of the spool valve shaft. The force acting on the cylinder piston will be the product of the cylinder piston area and the pressure difference between the high pressure and the exhaust. This force can be much larger than the force which moves the spool valve shaft.

Control elements

There are many forms of final control element. Thus, for example, in controlling the thickness of a rolled product in a manufacturing process the actuator may be a motor which rotates a screw which in turn moves rollers closer or further apart and so changes the thickness of the product (Fig. 6.8). In controlling the temperature of a bath of liquid the final control element may be an electric heater, switched on or off by a two-step controller. On a production line for packaged products the actuator may just operate some lever mechanism which knocks the package off the line if it is not up to the correct weight.

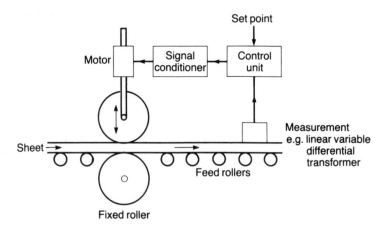

Fig. 6.8 Control of the thickness of a product

Because there are so many applications involving the control of the flow of fluids, the *pneumatic control valve* is a very widely used final control element. Figure 6.9 shows a cross-section of such a valve, complete with its actuator. The pneumatic or hydraulic pressure change in the actuator causes the diaphragm to move and so consequently the valve stem. The result of this motion is a movement of the inner-valve plug within the valve body. The plug restricts the fluid flow and so its position determines the flow rate.

There are many forms of valve body and plug. Figure 6.10 shows some of them. The term *single-seated* is used for a valve where there is just one path for the fluid through the valve and so just one plug is needed to control the flow. The term

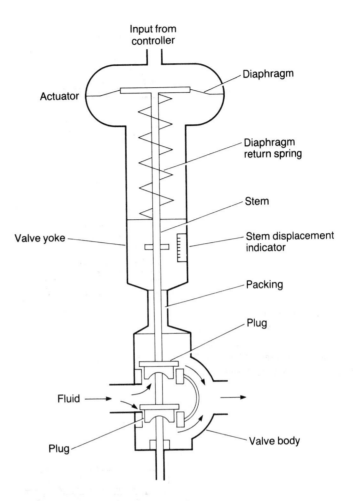

Input from
controller

Diaphragm

Actuator

Diaphragm
return spring

Stem

Valve yoke

Stem displacement
indicator

Packing

Plug

Fluid

Valve body

Plug

Fig. 6.9   Diaphragm-operated control
valve

double-seated is used for a valve where the fluid on entering
the valve splits into two streams, each stream passing through
an orifice controlled by a plug. There are thus two plugs with
such a valve. A single-seated valve has the advantage that it
can be closed more tightly than a double-seated one but the
disadvantage that the force on the plug due to the flow is much
higher and so the diaphragm in the actuator has to exert
considerably higher forces on the stem. This can result in
problems in accurately positioning the plug. For this reason
double-seated valves are more widely used when proportional
or three mode controllers are used.

The shape of the plug determines the relationship between
the stem movement and the effect on the flow rate. Figure
6.11 shows three commonly used types and Fig. 6.12 how the
percentage by which the volumetric rate of flow is related to
the percentage displacement of the valve stem. With the
quick-opening type a large change in flow rate occurs for a
small movement in the valve stem. Such a plug is used in two-

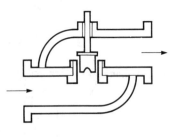

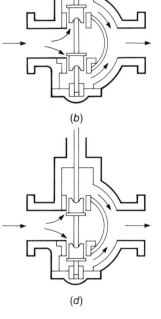
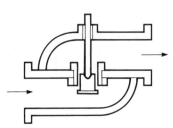

(a)

(b)

(c)

(d)

Fig. 6.10 Types of control valves,
(a) single-seated normally open,
(b) double-seated normally open,
(c) single-seated normally closed,
(d) double-seated normally closed

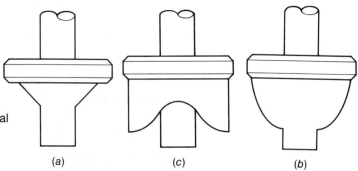

Fig. 6.11 Plug shapes, (a) quick
opening, (b) linear contoured, (c) equal
percentage

(a)

(c)

(b)

step control, i.e. on/off situations. With the *linear contoured*
type the change in flow rate is proportional to the change in
displacement of the valve stem, i.e.

Change in flow rate = k(change in stem displacement)

where k is a constant. If Q is the flow rate at a valve stem
displacement S and Q_{max} is the maximum flow rate at
maximum stem displacement S_{max}, then this means

$$\frac{Q}{Q_{max}} = \frac{S}{S_{max}}$$

or percentage change in the flow rate equals the percentage

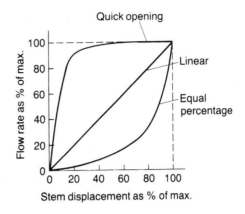

Fig. 6.12 Flow characteristics with different plugs

change in the stem displacement. With the *equal percentage* type the change in the rate of flow depends on the flow rate already occurring and the change in stem displacement, i.e.

$$\frac{\text{change in flow rate}}{\text{flow rate } Q} = k(\text{change in stem displacement } \Delta S)$$

Thus if the flow rate is Q then a 1% change in the stem displacement will give a 1% change in the value of Q. This means that if the flow rate was low the change would be smaller than if it was high when the change was made. The term equal percentage is used because the change in flow rate as a percentage of what the value was when the change was made is proportional to the percentage change in stem displacement. This is a typical property of an exponential relationship. Hence the equation can be written as

$$Q = Q_{min}e^{k\Delta s}$$

or alternatively

$$\frac{Q}{Q_{min}} = \left(\frac{Q_{max}}{Q_{min}}\right)^{S/S_{max}}$$

The relationships between the flow rate and the stem displacement is the inherent characteristic of a valve. It is only realised in practice if the pressure losses in the rest of the pipework, etc. are negligible compared with the pressure drop across the valve itself. If there are large pressure drops in the pipework so that, for example, less than half the pressure drop occurs across the valve then the linear characteristic of the valve can become almost a quick-opening characteristic. The linear characteristic is thus widely used when most of the system pressure is dropped across the valve. The effect of large pressure drops in the pipework with an equal percentage valve is to make it more like a linear characteristic. For this reason a valve with such a characteristic is widely used when only a small proportion of the system pressure is dropped across the valve.

The term *control valve sizing* is used for the procedure of determining the correct size of valve body. The equation relating the flow of liquid through the wide open valve to its size is

$$Q = A_v\sqrt{(\Delta P/\rho)}$$

where Q is volume rate of flow through the fully open valve in m^3/s, ΔP is the pressure drop across the valve in Pa, ρ is the density in kg/m^3 and A_v is the valve flow coefficient. This coefficient depends on the size of the valve and tables are available relating its value to size. This equation is sometimes expressed as

$$Q = 2.37 \times 10^{-5} C_v\sqrt{(\Delta P/\rho)}$$

where C_v is the valve flow coefficient. Alternatively

$$Q = 0.75 \times 10^{-6} C_v\sqrt{(\Delta P/G)}$$

where G is the specific gravity or relative density. These last two forms of the equation derive from its original specification in terms of US gallons. Table 6.1 shows some typical values of A_v, C_v, and valve size.

Table 6.1 Flow coefficients and valve sizes

Valve size (mm)	480	640	800	960	1260	1600	1920	2560
C_v	8	14	22	30	50	75	110	200
$A_v \times 10^{-5}$	19	33	52	71	119	178	261	474

Example 1

An actuator has a stem movement which at full travel is 30 mm. It is mounted with a linear valve which has a minimum flow rate of $0\,m^3/s$ and a maximum flow rate of $40\,m^3/s$. What will be the flow rate when the stem movement is (*a*) 10 mm, (*b*) 20 mm?

Answer

Since the percentage flow rate is the same as the percentage stem displacement then:
(*a*) percentage stem displacement = 33% and so percentage flow rate is 33%, i.e. $13\,m^3/s$.
(*b*) percentage stem displacement = 67% and so percentage flow rate is 67%, i.e. $27\,m^3/s$.

Example 2

An actuator has a stem movement which at full travel is 30 mm. It is mounted with an equal percentage valve which has a minimum flow rate of $2\,m^3/s$ and a maximum flow rate of $24\,m^3/s$. What will be the flow rate when the stem movement is (*a*) 10 mm, (*b*) 20 mm?

Answer

Using the equation given earlier

$$\frac{Q}{Q_{min}} = \left(\frac{Q_{max}}{Q_{min}}\right)^{S/S_{max}}$$

(a) $Q = 2 \times (24/2)^{10/3} = 4.6\,\text{m}^3/\text{s}$

(b) $Q = 2 \times (24/2)^{20/30} = 10.5\,\text{m}^3/\text{s}$

Example 3

What is the valve size required for a valve that is required to control the flow of water when the maximum flow required is $0.012\,\text{m}^3/\text{s}$ and the permissible pressure drop across the valve at this flow rate is $300\,\text{kPa}$? The density of water is $1000\,\text{kg/m}^3$.

Answer

Using the equation given above

$$Q = A_v\sqrt{(\Delta P/\rho)}$$

$$A_v = Q\sqrt{(\rho/\Delta P)} = 0.012\sqrt{(1000/300 \times 1000)} = 69.3 \times 10^{-5}$$

The valve size required is thus, using Table 6.1, $960\,\text{mm}$.

Problems

1   Explain the purpose of the actuator in a control system.
2   Describe the operation of a diaphragm pneumatic actuator.
3   Explain the difference in operation and characteristics of single-seated and double seated valves.
4   Explain how the inherent characteristics of (a) quick opening, (b) linear, (c) equal percentage valves differ and the circumstances under which each is likely to be used.
5   An actuator has a stem movement which at full travel is $40\,\text{mm}$. It is mounted with a linear valve which has a minimum flow rate of $0\,\text{m}^3/\text{s}$ and a maximum flow rate of $0.20\,\text{m}^3/\text{s}$. What will be the flow rate when the stem movement is (a) $10\,\text{mm}$, (b) $20\,\text{mm}$?
6   An actuator has a stem movement which at full travel is $40\,\text{mm}$. It is mounted with an equal percentage valve which has a minimum flow rate of $0.2\,\text{m}^3/\text{s}$ and a maximum flow rate of $4.0\,\text{m}^3/\text{s}$. What will be the flow rate when the stem movement is (a) $10\,\text{mm}$, (b) $20\,\text{mm}$?
7   What is the valve size required for a valve that is required to control the flow of water when the maximum flow required is $0.002\,\text{m}^3/\text{s}$ and the permissible pressure drop across the valve at this flowrate is $100\,\text{kPa}$? The density of water is $1000\,\text{kg/m}^3$.

7 Data display

The range of data-presentation elements

There is a very wide range of elements used for the presentation of data. They can be broadly classified into two groups: indicators and recorders. *Indicators* give an instant visual indication of the process variable while *recorders* record the output signal over a period of time and give automatically a permanent record. The recorder will be the most appropriate choice if the event is high speed or transient and not able to be followed by an observer, there are large amounts of data, or it is essential to have a record of the data.

Both indicators and recorders can be subdivided into two groups of devices, *analogue* and *digital*. An example of an analogue indicator is a meter which has a pointer moving across a scale, while a digital meter would be just a display of a series of numbers. An example of an analogue recorder is a chart recorder which has a pen moving across a moving sheet of paper, while a digital recorder has the output recorded on a sheet of paper as a sequence of numbers.

The moving-coil meter

The *moving-coil meter* is an analogue indicator with a pointer moving across a scale. The amount of movement of the pointer across the scale is related to the input to the meter. The basic instrument movement is a microammeter with shunts, multipliers and rectifiers being used to convert it to other ranges and measurements. Essentially the meter movement consists of a coil situated in a constant radial magnetic field, i.e. a magnetic field which is always at right-angles to the sides of the coil no matter what angle the coil has rotated through (Fig. 7.1). When a current passes through the coil, forces act on the coil sides and so there is a deflecting torque. This deflecting torque is proportional to the current and causes the coil to rotate. There is an opposing torque generated by springs, the size of this torque depending on the angle through

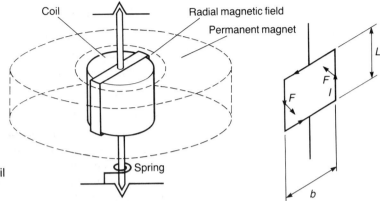

Fig. 7.1 The basis of the moving-coil meter

which the coil rotates. The result is that equilibrium is reached and a steady deflection obtained when the torque due to the current is exactly balanced by the torque to the springs. Hence the angular movement of the coil is proportional to the current.

Consider a coil carrying a current I with vertical sides of length L and horizontal sides length b in a magnetic field which has a uniform flux density B which is always at right-angles to the coil. The forces acting on the horizontal sides are in opposite directions and since each side carries the same current these forces cause no motion of the coil, however the forces acting on the vertical sides are also in opposite directions and the same size but they cause rotation. The force F acting on a vertical side is

$$F = BIL$$

The torque about the central vertical axis of the coil is thus

$$\text{torque} = BIL \times \tfrac{1}{2}b + BIL \times \tfrac{1}{2}b = BILb$$

Lb is the area A of the loop. Hence

$$\text{torque} = BIA$$

This is the torque on a single turn of the coil. If there are N turns on the coil then

$$\text{torque} = NBIA$$

Since, for a particular galvanometer NB and A will be constant,

$$\text{torque} = K_c I$$

where K_c is a constant for that galvanometer.

This torque causes the coil to rotate against springs. The opposing torque developed by the springs is proportional to the angular deflection θ.

torque due to springs $= K_s\theta$

where K_s is constant. The coil thus rotates until the torque developed by the springs cancels that produced by the current through the coil. Then

$$K_c I = K_s \theta$$

The angular deflection θ is proportional to the current I. A more detailed discussion of the principles of the moving-coil system is given later in this chapter.

Moving-coil meters generally have resistances of the order of a hundred ohms. The accuracy of such a meter depends on a number of factors. Among them are temperature, the presence nearby of magnetic fields or ferrous materials, the way the meter is mounted, bearing friction, inaccuracies in scale marking during manufacturer, etc. In addition there are human errors involved in reading the meter. Such errors arise as a result of parallax when the position of the pointer is read from an angle other than directly at right-angles to the scale and also from errors in interpolating between scale markings. The overall result is that accuracies are generally of the order of $\pm 0.1\%$ to $\pm 5\%$. The time taken for a moving-coil meter to reach a steady deflection is typically in the region of a few seconds.

Example 1

On what factors does the current sensitivity of a moving-coil galvanometer depend?

Answer

The current sensitivity is θ/I and so, using the equations developed above, is K_c/K_I. Since K_c is NBA, the sensitivity is thus NBA/K_s. The larger the number of coil turns, the larger the flux density and the larger the coil area the greater the sensitivity. The smaller the spring constant K_s the greater the sensitivity.

The digital meter

The *digital meter* gives its reading in the form of a sequence of digits. Such a form of display eliminates parallax and interpolation errors and can give accuracies as high as $\pm 0.005\%$ and a resistance of the order of $10\,\text{M}\Omega$. The digital meter is based on the digital voltmeter, of which there are a number of forms. All the forms can however be considered to be an analogue-to-digital converter connected to a counter (Fig. 7.2), with the differences being the different forms of digital-to-analogue converters.

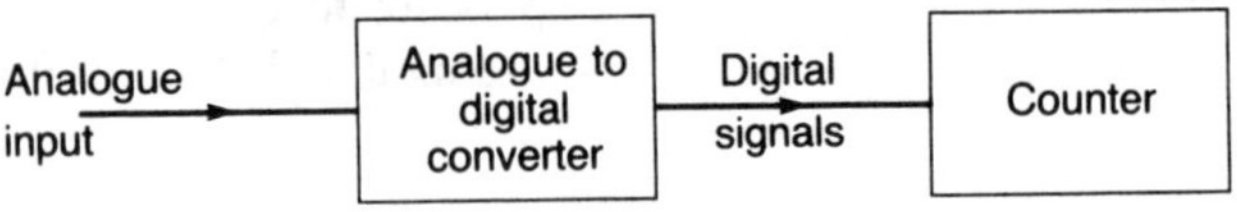

Fig. 7.2 Digital voltmeter principle

Analogue input → Analogue to digital converter → Digital signals → Counter

One form of instrument is the *successive approximations digital voltmeter* (Fig. 7.3). This type of instrument is capable of sampling a voltage 1000 times per second and so is widely used for fast-changing voltages. The voltage of the instrument is divided into increments which diminish in size, e.g. 1/2, 1/4, 1/8, 1/16, 1/32 of some reference voltage. The counter switches each of these increments in turn for comparison with the voltage being measured. The voltage to be measured is compared with these increments of this reference voltage which are added together until they equal it. For example, the first comparison will be with the 1/2 reference voltage. If the voltage is bigger than this then the 1/2 reference voltage is retained and it is counted. Then the next increment, the 1/4, is added to it and the sum compared with the voltage. If the voltage is less than this then the 1/4 is discarded and the 1/8 added to the 1/2 and another comparison made. The process is continued until the sum of the increments equals the voltage. Each increment of voltage logs up digits and so the total number is the voltage.

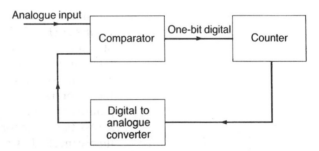

Fig. 7.3 Successive approximations digital voltmeter

The *ramp type digital voltmeter* is essentially a voltage-to-time converter with the display indicating time (Fig. 7.4). At the start of the measurement a ramp voltage starts (Fig. 7.5). This voltage is continuously compared with the voltage being measured. When the two are equal a pulse is generated which opens a gate. The ramp voltage continues until it reaches zero. Another pulse is generated which then closes the gate. During

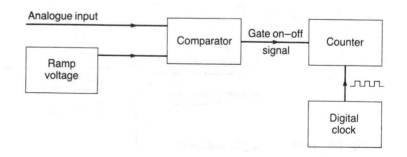

Fig. 7.4 Ramp-type digital voltmeter

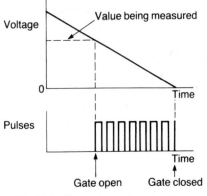

Fig. 7.5 Ramp-type digital voltmeter

the time the gate is open time pulses are counted. The result is that the counted number of pulses is a measure of the voltage being measured. The reading given by such an instrument is a sample of the voltage since it is just the value that happened to occur at the instant its value coincided with the ramp voltage value.

Another form of instrument is the *voltage-to-frequency digital voltmeter* (Fig. 7.6). In this method a signal is generated whose frequency is related to the size of the input voltage. The number of pulses in this signal is then counted over some time interval and so is a measure of the voltage. The time interval often used is that of 1 cycle of the mains alternating supply. This has the advantage of eliminating mains noise from the signal.

Fig. 7.6 Voltage-to-frequency digital voltmeter

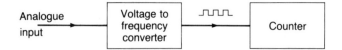

Another form of instrument is the *dual-slope digital voltmeter* (Fig. 7.7). Essentially this involves a capacitor, really an integrator, being charged by the input voltage during a time interval which equals that of 1 cycle of the mains frequency. Then the switch disconnects the input voltage and switches to the reference voltage. The reference voltage is of opposite polarity to the voltage being measured. The potential difference across the capacitor which resulted from the initial charging is then cancelled at a steady rate. Then the time taken for it to reach zero is measured. The measurement is by means of counting the number of pulses produced by some 'clock' during that time. This method has the advantage of eliminating mains noise from the signal. The method cannot however be used for fast-changing voltages since readings cannot be taken at a rate faster than that of 1 cycle of the mains frequency.

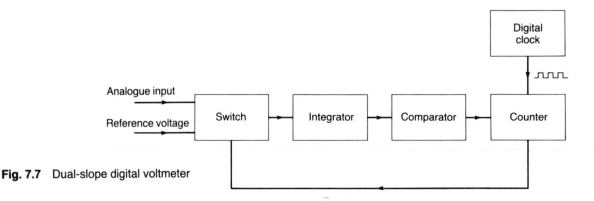

Fig. 7.7 Dual-slope digital voltmeter

Example 2

A digital voltmeter specification includes the statement: sample rate approximately 5 readings per second. What is the significance of this?

Answer

The sample rate of 5 readings per second means that every 0.2 s the input voltage is sampled. It is the time taken for the instrument to process the signal and give a reading. Thus if the input voltage is changing at a rate which results in significant changes during 0.2 s then the voltmeter reading can be in error.

Alarm indicators

A wide variety of alarm systems are used. Commonly met ones are:

1   Temperature alarms which respond when the temperature reaches a particular value or falls to some other value. These may be based on the use of a resistance element or a thermocouple.
2   Current alarms which respond when the current reaches a particular value or falls to some other value.
3   Voltage alarms which respond when the voltage reaches a particular value of falls to some other value.
4   Weight alarms which respond when the weight in a container reaches a particular value or falls to some other value. These generally use load cells with electrical resistance strain gauges.

Alarm indicators take an analogue input from some transducer (possibly via a signal conditioner) and turn it into an on–off signal for some indicator. Figure 7.8 shows the basic form of alarm systems. The input is compared with the alarm level set point. When the set point is exceeded the logic unit triggers the switching unit which switches on, or off, an indicator. The indicator can take a variety of forms, e.g. a bell, a horn, a klaxon, a coloured light, a flashing light, a backlighted display (the light comes on behind a message on a screen).

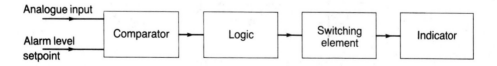

Fig. 7.8 An alarm system

Analogue chart recorders

With analogue chart recorders the input signal is translated into some position on a chart and a marking mechanism marks the point. By either moving the marking mechanism or the chart in a controlled way with time an analogue record can be obtained of how the input signal varied with time. The two main forms of chart recorder are the galvanometric type and the potentiometric type.

Galvanometric recorders

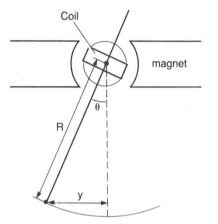

Fig. 7.9 Galvanometric type of chart recorder

The *galvanometric type* of chart recorder (Fig. 7.9) works on the same principle as the moving-coil meter movement described earlier in this chapter. The main difference is that usually the torque opposing that generated by the current through the coil is produced by the twisting of the suspension of the coil rather than springs. A pointer may be attached to the suspension with a pen at its end. An ink trace is then produced on the chart.

If R is the length of the pointer and θ the angular deflection of the coil, then the displacement y of the pen is

$$y = R \sin \theta$$

Since the angle θ is proportional to the current through the coil, then the relationship between the current and the displacement is non-linear. If the angular deflections are restricted to less than $\pm 10°$ then the error due to this non-linearity is less than 0.5%. A greater problem is however the fact that the pen moves in an arc rather than a straight line. Thus curvilinear paper (Fig. 7.10) is used for the plotting. However there are difficulties in interpolation for points between the curved lines.

An alternative form of 'pen' chart recorder which leads to rectilinear charts rather than curvilinear charts is the *knife-edge recorder* (Fig. 7.11). One version of this uses heat-sensitive paper which moves over a knife edge and a heated stylus instead of an ink pen. The paper may be impregnated with a chemical that shows a marked colour change when heated by contact with the stylus or the stylus burns away temperature-sensitive outer layers which coat the paper. The use of the knife edge avoids the curved trace but the non-linear relationship between θ and displacement y still exists. The length of trace y produced on the paper by a deflection θ is

$$y = R \tan \theta$$

The non-linearity error is slightly greater than for the pen form of recorder. If deflections are restricted to less than $\pm 10°$ the error is less than 1%.

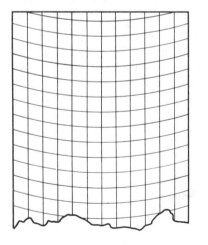

Fig. 7.10 Curvilinear chart paper

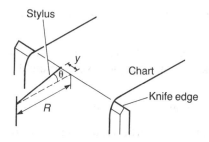

Fig. 7.11 Knife-edge recorder

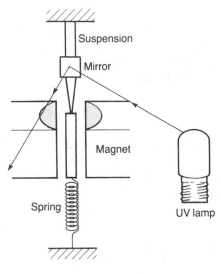

Fig. 7.12 Ultraviolet galvanometric recorder

Usually with the pen or knife-edge form of galvanometric recorder there is some electronic amplification of the input signal. This leads to sensitivities which are generally of the order of 1 centimetre of pen displacement per millivolt, input resistances of about $10\,k\Omega$, a bandwidth from d.c. to about $50\,Hz$ and accuracies of the order of $\pm 2\%$ of the full-scale deflection.

There are a number of ways by which the movement of the coil may be transformed into a trace on a chart. An alternative to the pen form is the *ultraviolet recorder* which involves a small mirror being attached to the suspension (Fig. 7.12). A beam of ultraviolet light is directed at the mirror and thus when the coil rotates the reflected beam is swept across the chart. The chart uses photosensitive paper and so a trace is produced when it is developed.

The use of an 'optical pointer' rather than a strip of material enables longer pointer lengths to be used and so greater sensitivity. With a bandwidth of d.c. to about $50\,Hz$ the sensitivity is typically about $5\,cm/mV$, the coil having a resistance of about $80\,\Omega$. Higher bandwidth instruments have lower sensitivities, e.g. a bandwidth extending up to $5\,kHz$ with a sensitivity of about $0.0015\,cm/mV$ and a coil resistance of about $40\,\Omega$. The limiting frequency for this type of instrument is about $13\,kHz$. Typically the accuracy is about $\pm 2\%$ of the full-scale deflection. Since optical pointers can cross each other without interference it is quite common to have 6, 12 or 25 galvanometer mountings side by side in the same magnet block and so enable simultaneous recordings of many variables to be made.

Dynamic behaviour of galvanometric recorders

When the current through a galvanometer coil suddenly changes from zero to some value I, a so-called step input, then the coil experiences a torque which is proportional to the current.

Torque due to current $I = K_c I$

where K_c is a constant for the coil and magnet arrangement. K_c has the value NAB, where N is the number of turns on the coil, A the cross-sectional coil area and B the flux density at right-angles to the coil (see earlier this chapter). The coil will begin to rotate and this rotation is opposed by a torque due to a spring or the twisting of the coil suspension. This torque is proportional to the angle θ through which the coil has moved.

Opposing torque $= K_s \theta$

where K_s is a constant related to the spring or suspension used. The net torque acting on the coil is thus

net torque = torque due to current − opposing torque
$$= K_c I - K_s \theta$$

Just as with linear motion where a net force F gives a linear acceleration a $(F = ma)$, so this net torque gives an angular acceleration.

Net torque = moment of inertia J × angular acceleration

For angular motion the moment of inertia is the equivalent of mass in linear motion. Hence

$$J \times \text{angular acceleration} = K_c I - K_s \theta$$

The current through the coil originates from some transducer/signal conditioner arrangement. The situation can be considered to be effectively the Thévenin equivalent circuit shown in Fig. 7.13. The current I is thus

$$I = \frac{V_t}{R_t + R_r}$$

where V_t is the Thévenin equivalent voltage output from the transducer/signal conditioner arrangement, R_t its effective resistance and R_r the effective resistance of the recorder circuit in which the galvanometer coil is.

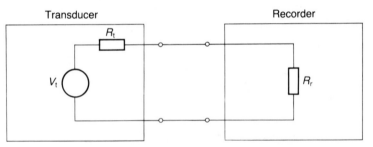

Fig. 7.13 Transducer connected to the chart recorder

This initial current causes the coil to rotate. But the coil is rotating in a magnetic field and so there is an induced e.m.f. This induced e.m.f. is proportional to the angular velocity of the coil, the constant of proportionality being the same as the constant K_c relating the current in the coil to the torque. Thus the voltage in the circuit is

$$V_t - (K_c \times \text{angular velocity})$$

Hence the circuit current I is, during the movement of the galvanometer coil,

$$I = \frac{V_t - (K_c \times \text{angular velocity})}{R_t + R_r}$$

Hence substituting for I in the angular acceleration equation

$J \times$ angular acceleration

$$K_c \left[\frac{V_t - (K_c \times \text{angular velocity})}{R_t + R_r} \right] - K_s \theta$$

The angular velocity is the rate at which the angle changes with time and can be written as $d\theta/dt$. The angular acceleration is the rate at which the angular velocity changes with time and can be written as $d^2\theta/dt^2$. Hence

$$\frac{d^2\theta}{dt^2} + \frac{K_c^2}{J(R_t + R_r)} \frac{d\theta}{dt} + \frac{K_s\theta}{J} = \frac{K_c V_t}{J(R_t + R_r)}$$

This is a second-order differential equation and describes the motion of the galvanometer coil.

When the recorder galvanometer reaches the steady reading both $d^2\theta/dt^2$ and $d\theta/dt$ are zero. Hence

$$\frac{K_s\theta}{J} = \frac{K_c V_t}{J(R_t + R_r)}$$

The steady state voltage sensitivity, is thus

$$\text{voltage sensitivity} = \frac{\theta}{V_t} = \frac{K_c}{K_s(R_t + R_r)}$$

The differential equation describes the behaviour of a system which has a natural frequency of oscillation $\omega_n = 2\pi f_n$ given by

$$\omega_n = \surd(K_s/J)$$

and is damped with a damping factor of

$$\text{damping factor} = \frac{K_c^2}{2(K_s J)^{\frac{1}{2}}(R_t + R_r)}$$

The damping factor involves the effective resistance of the transducer. Thus the damping can be altered by adding resistors in series or parallel with the transducer.

Figure 7.14 shows how the response of the galvanometer depends on the damping factor when there is a step current input. The *damping factor* expresses the damping as a fraction of that which gives critical damping. Critical damping occurs when the galvanometer coil responds to the step input by changing to the steady deflection without any overshoot in the minimum of time. In Fig. 7.14 it is the deflection indicated for the damping factor of 1.0. At damping factors which are less than 1.0 the deflection overshoots the steady value before settling back to it. With a damping factor of 0.4 the percentage overshoot is about 25%, at 0.7 it is about 5%. With damping factors greater than 1.0 the coil just takes a long time to attain the steady value.

The response of the galvanometer to different frequency transducer inputs depends on the value of the natural

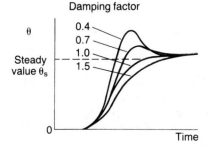

Fig. 7.14 Response of the galvanometer to a step current input

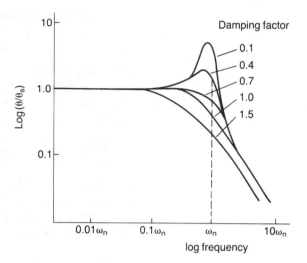

Fig. 7.15 Frequency response of a
recorder galvanometer

frequency and the damping. Figure 7.15 shows the frequency
response. The frequency band for which the steady deflection
value θ_s is attained depends on the damping. The maximum
frequency for which this occurs is when the damping factor is
of the order of 0.7 and is almost the natural frequency.

Galvanometers for use at high frequencies thus require a
damping factor of about 0.7 and a high natural frequency. One
way by which the natural frequency can be increased is by
reducing the moment of inertia J of the coil. This can be
achieved by using a slim coil, i.e. a small breadth. However
reducing the moment of inertia also affects the damping
factor, a decrease in moment of inertia increasing the damping
factor.

Another way by which the natural frequency can be
increased is by increasing K_s. This however also affects the
sensitivity of the galvanometer. Thus galvanometers designed
for use at high frequencies tend to have low sensitivities.

Example 3

How is the damping factor and steady state voltage sensitivity of a
recorder galvanometer affected by a resistance of $100\,\Omega$ being
included in series with a transducer, if the damping factor is 0.6
without it? The recorder has a resistance of $60\,\Omega$ and the transducer a
resistance of $160\,\Omega$.

Answer

Using the equation for the damping factor given above

$$\text{damping factor} = \frac{K_c^2}{2(K_s J)^{\frac{1}{2}}(R_t + R_r)}$$

Hence, without the extra $100\,\Omega$,

$$0.6 = \frac{K_c^2}{2(K_sJ)^{\frac{1}{2}}(160 + 60)}$$

With the extra $100\,\Omega$,

$$\text{damping factor} = \frac{K_c^2}{2(K_sJ)^{\frac{1}{2}}(260 + 60)}$$

Hence, dividing these two equations,

$$\frac{\text{damping factor}}{0.6} = \frac{160 + 60}{260 + 60}$$

The new damping factor is thus 0.4. The result of this is that the galvanometer will give considerably more overshoot.

The steady state voltage sensitivity is given by

$$\text{sensitivity} = \frac{\theta}{V_t} = \frac{K_c}{K_s(R_t + R_r)}$$

Without the extra resistance,

$$\text{sensitivity without } R = \frac{K_c}{K_s(160 + 60)}$$

With the extra resistance

$$\text{sensitivity with } R = \frac{K_c}{K_s(260 + 60)}$$

Hence, dividing these two equations,

$$\frac{\text{sensitivity with } R}{\text{sensitivity without } R} = \frac{160 + 60}{260 + 60}$$

The voltage sensitivity is decreased by a factor of 0.7.

Example 4

How is the damping factor and the steady state voltage sensitivity of a moving coil galvanometer affected by a $100\,\Omega$ resistance being connected (a) in series with the transducer and (b) in parallel with it? The galvanometer has a resistance of $80\,\Omega$ and the transducer a resistance of $100\,\Omega$.

Answer

The damping factor is given by

$$\text{damping factor} = \frac{K_c^2}{2(K_sJ)^{\frac{1}{2}}(R_t + R_r)}$$

(a) Hence for the resistance in series

$$\frac{\text{new damping factor}}{\text{initial damping factor}} = \frac{R_t + R_r}{R + R_t + R_r} = \frac{180}{280}$$

The new damping factor is 0.64 times the initial damping factor.

The steady state voltage sensitivity is given by

$$\text{sensitivity} = \frac{\theta}{V_t} = \frac{K_c}{K_s(R_t + R_r)}$$

Hence for the resistance in series

$$\frac{\text{new sensitivity}}{\text{initial sensitivity}} = \frac{R_t + R_r}{R + R_t + R_r} = \frac{180}{280}$$

The new sensitivity is 0.64 times the initial sensitivity.

(b) For the resistance in parallel the effective value of the transducer resistance in parallel with the added resistor is given by

$$\frac{1}{R} = \frac{1}{100} + \frac{1}{100}$$

and so is $50\,\Omega$. Hence

$$\frac{\text{new damping factor}}{\text{initial damping factor}} = \frac{280}{130}$$

The new damping factor is 2.2 times the initial damping factor.

In considering the voltage sensitivity account has to be taken of not only the change in the effective resistance but also the voltage applied to the recorder. The Thévenin equivalent voltage for the transducer is reduced by a factor of $R_p/(R_p + R_t)$, where R_p is the resistance inserted in parallel to the transducer. Thus in this case the Thévenin equivalent voltage is reduced by a factor of $100/(100 + 100)$, i.e. by a half. Hence, since

$$\text{voltage sensitivity} = \frac{\theta}{V_t} = \frac{K_c}{K_s(R_t + R_r)}$$

with the parallel resistor,

$$\frac{\theta}{0.5\,V_t} = \frac{K_c}{K_s(50 + 80)}$$

and so the new sensitivity is

$$\text{new sensitivity} = \frac{\theta}{V_t} = \frac{0.5\,K_c}{K_s \times 130}$$

Since the initial sensitivity was

$$\text{initial sensitivity} = \frac{K_c}{K_s(100 + 80)}$$

then the new sensitivity is $0.5 \times 180/130 = 0.69$ times the initial sensitivity.

Example 5

A moving-coil galvanometer movement has a coil of 100 turns, a cross-sectional area of $1.0 \times 10^{-4}\,\text{m}^2$, a moment of inertia of $2.4 \times 10^{-5}\,\text{kg}\,\text{m}^2$, a resistance of $60\,\Omega$ and a magnetic field of flux density $80\,\text{T}$. The restoring spring applies a torque of $1.2 \times 10^{-2}\,\text{N}$ m per radian of coil rotation. What will be (a) the steady state voltage sensitivity, (b) the damping ratio, (c) the natural frequency, when the signal source connected to the movement has a resistance of $240\,\Omega$?

Answer

(*a*) Using the equation for steady state voltage sensitivity given earlier in this chapter

$$\text{sensitivity} = \frac{\theta}{V_i} = \frac{K_c}{K_s(R_t + R_r)}$$

K_c has the value NAB, where N is the number of turns, A the coil area and B the flux density. Hence

$$\text{sensitivity} = \frac{100 \times 1.0 \times 10^{-4} \times 80}{1.2 \times 10^{-2}(240 + 60)} = 0.22 \, \text{rad/V}$$

(*b*) Using the equation for the damping ratio given earlier in this chapter

$$\text{damping factor} = \frac{K_c^2}{2(K_s J)^{\frac{1}{2}}(R_t + R_r)}$$

$$= \frac{(100 \times 1.0 \times 10^{-4} \times 80)^2}{2(1.2 \times 1^{-2} \times 2.4 \times 10^{-5})^{\frac{1}{2}}(240 + 60)} = 7.3$$

(*c*) Using the equation for natural frequency given earlier in this chapter

$$\begin{aligned}\omega_n &= \sqrt{(K_s/J)} \\ &= \sqrt{(1.2 \times 10^{-2}/2.4 \times 10^{-5})} = 22.4 \, \text{Hz}\end{aligned}$$

Hence since $\omega = 2\pi f$, the frequency is 3.6 Hz.

Potentiometric recorders

Figure 7.16 illustrates the general principles of the potentiometric recorder. Such a recorder is sometimes referred to as a *closed-loop recorder* or a *closed-loop servo recorder*. The position of the pen is monitored by means of a slider which

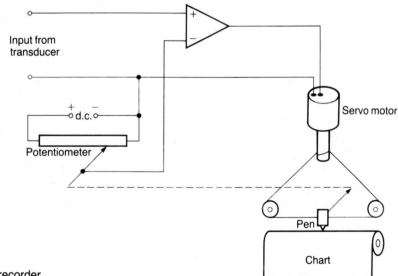

Fig. 7.16 Potentiometric recorder

LIVERPOOL JOHN MOORES UNIVERSITY
LEARNING SERVICES

moves along a linear potentiometer. The position of the slider determines the potential applied to an operational amplifier. The amplifier subtracts the measurement signal from the signal from the transducer. The output from the amplifier is thus a signal related to the difference between the pen and transducer signals. This signal is used to operate a servo motor which in turn controls the movement of the pen across the chart. The pen thus ends up moving to a position where the result is no difference between the pen and transducer signals.

Potentiometric recorders typically have high input resistances, higher accuracies (about $\pm 0.1\%$ of full-scale reading) than galvanometric recorders but considerably slower response times. Response times are typically of the order of 1–2 s and so the bandwidth is only d.c. to 1 or 2 Hz. They can thus only really be used for d.c. or slowly changing signals. Because of friction there is a minimum current required to get the motor operating. There is thus some error due to the recorder not responding to a small transducer signal. This error is known as the *dead band*. Typically it is about $\pm 0.3\%$ of the range of the instrument. Thus for a range of 5 mV the dead band error amounts to ± 0.015 mV.

Cathode-ray oscilloscope

Figure 7.17 shows the basic features of the cathode-ray tube. Electrons are produced by the heating of the cathode. The number of these electrons which form the electron beam, i.e. its brilliance, is determined by a potential applied to an electrode, the modulator, immediately in front of the cathode. The electrons are accelerated down the tube by the potential difference between the cathode and the anode. An electron lens is used to focus the beam so that when it reaches the phosphor-coated screen it forms a small luminous spot. The focus is adjusted by changing the potential of the electrodes

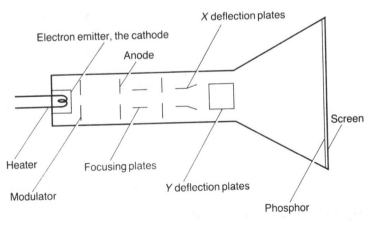

Fig. 7.17 Cathode-ray tube

relative to that of the earlier electrodes. The electron emission, modulator, anode and lens is known as the electron gun. The beam can be deflected in the Y direction by a potential difference applied between the Y deflection plates. A potential difference between the X deflection plates will cause it to deflect in the X direction. The screen has a scale, a rectangular grid. The oscilloscope is an example of an instrument employing a cathode-ray tube, another being the TV monitor.

The *phosphor* used as the screen-coating material emits light when hit by electrons. The light takes a little time to build up when the electron beam first hits the spot and time to decay when the electron beam ceases to hit the spot. The time taken for the light output to fall to some specified value of its initial value is known as the *decay time* or *persistence*. A phosphor used widely with oscilloscopes is P31. It gives a yellowish green trace and a decay time to 0.1% of 32 ms. Where the oscilloscope is used with a camera for photographing high speed traces P11 is used. It gives a blue trace with a decay time to 0.1% of 20 ms.

A signal applied to the *Y deflection plates* causes the electron beam, and hence the spot on the screen, to move up or down in the vertical direction. A switched attenuator, or range selector, and amplifier are used between the incoming signal and the plates in order that different deflection sensitivities can be obtained. A general-purpose oscilloscope is likely to have sensitivities which vary between 5 mV per scale division to 20 V per scale division. The amplifier generally has an a.c. or d.c. mode of operation switch. The a.c. mode introduces a blocking capacitor in the input line to eliminate drift due to d.c. components. When the amplifier is in its a.c. mode its bandwidth typically extends from about 2 Hz to 10 MHz, when in its d.c. mode from d.c. to 10 MHz.

Figure 7.18 shows the type of response that is obtained when a step input, i.e. an abrupt change in input voltage, is applied to the Y deflection system. Some time elapses before the deflection indicates 100% of the applied voltage, with some overshoot occurring before the steady 100% value occurs. The term *rise time* is used for the time taken for the deflection to go from 10% to 90% of its steady deflection. In general, the product of the rise time and bandwidth is a constant in the range 0.25–0.5. For minimum overshoot the optimum value is 0.35. Thus a bandwidth of 10 mHz would mean a rise time of about 35×10^{-9} s (35 ns).

The effect of the oscilloscope rise time on a measurement is to give a false reading of the real rise time of the input signal. Thus if the signal has a rise time t_s and the oscilloscope a rise time t_o then the indicated signal has a rise time t_i given by

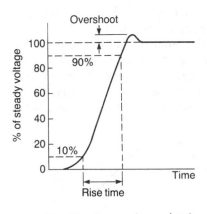

Fig. 7.18 Rise time and overshoot

$$t_i^2 = t_s^2 + t_o^2$$

The error of the indicated rise time is the difference between the indicated rise time and the actual signal rise time. With the signal rise time about twice the oscilloscope rise time the error is about 10%. For errors less than 2% the signal rise time has to be five times that of the oscilloscope.

The *X deflection plates* are generally used with an internally generated signal, a voltage with a sawtooth waveform (Fig. 7.19). This sweeps the luminous spot on the screen from left to right at a constant velocity with a very rapid return, i.e. flyback. This return is too fast to leave a trace on the screen. The constant velocity movement from left to right means that the distance moved in the X direction is proportional to the time elapsed. Hence the sawtooth waveform gives a horizontal time axis, i.e. a *time base*. Typically an oscilloscope will have a range of time bases from about 1s per scale division to $0.2\,\mu s$ per scale division.

For an input signal to give rise to a steady trace on the screen it is necessary to synchronise the timebase and the input signal. For this purpose a *trigger circuit* is used. The trigger circuit can be adjusted so that it responds to a particular voltage level and also whether the voltage is increasing or decreasing. This means that for a periodic signal input the trigger circuit responds to particular points in its cycle (Fig. 7.20). Pulses are produced which in turn trigger the time base into action. The time-base sweep across the screen thus always starts at the same point on the input signal. The result is that successive scans of the input signal are superimposed.

Double beam oscilloscopes enable two separate traces to be observed simultaneously on the screen. One way this is achieved is by having two independent electron gun assemblies

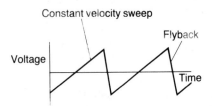

Fig. 7.19 Sawtooth waveform

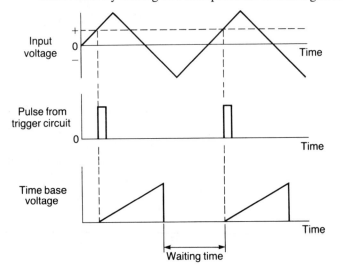

Fig. 7.20 Triggering

and so two electron beams. Each beam has its own Y deflection plates but a common set of X deflection plates and so a common time base. A cheaper, more commonly used, version is obtained by using a single electron gun and switching the Y deflection plates from one input signal to the other each time the time base is triggered. This is known as the *alternate mode*. This has the disadvantage that the two events indicated by the two traces are not occurring at exactly the same time. However, if the two events are cyclical this presents no problems. Another alternative is to sample the two inputs more frequently. This is known as the *chopped mode*. The frequency with which the deflection plates are switched from one signal to the other is typically about 150 kHz.

With *storage oscilloscopes* the trace produced by the Y deflection plates remains on the screen after the input signal has ceased, only being removed by a deliberate action of erasure. Figure 7.21 shows the basic features of a bistable storage tube. The tube has three electron guns. Two of the guns, called the flood electron guns, are on all the time and permanently flood the viewing area with low-velocity electrons. The viewing area consists of phosphor particles on a dielectric sheet, backed by a conducting layer. When the low-velocity flood gun electrons fall on a phosphor particle they charge it up. It thus becomes negatively charged and begins to repel further electrons. It thus reaches a stable charge value and no further electrons hit it. The phosphor is in a 'not glowing' state and remains in that condition. The writing electron gun emits high-velocity electrons. When it is on the electrons have sufficient velocity to overcome the negative charge on a phosphor particle which resulted from the flood guns. The velocity is high enough for the electrons to knock further electrons out of the phosphor particle. These electrons are gathered by the conducting layer which backs the phosphor-coated sheet. The result of the phosphor particles losing electrons is that they become positively charged. This charge remains, even when the writing gun stops emitting electrons. This is because the phosphor particle is being bombarded by flood electrons which are accelerated towards it by the positive charge. The acceleration is sufficient for secondary emission to continue. Thus the phosphor is in a 'glowing state' and remains in that condition.

Example 6

The Y deflection system of an oscilloscope has a bandwidth of d.c. to 15 MHz. What would be the expected rise time?

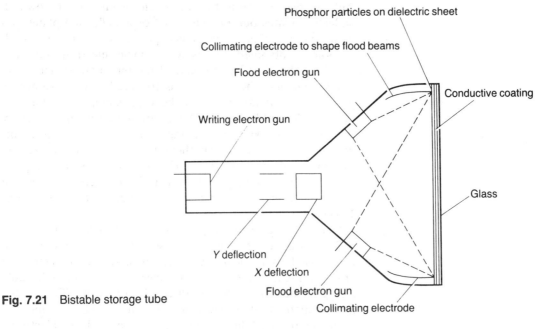

Fig. 7.21   Bistable storage tube

Answer

Using the equation given earlier in this chapter

$$\text{bandwidth} \times \text{rise time} \approx 0.35$$

then the rise time is about $0.35/(15 \times 10^6) = 23 \times 10^{-9}\,$s, i.e. 23 ns.

Monitors

A monitor is just a display device, involving a cathode-ray tube, which can display alphanumeric, graphic and pictorial data. With the cathode-ray tube in the form shown in Fig. 7.17 sawtooth signals are applied to both the X and the Y deflection plates (Fig. 7.22). The Y input moves the spot relatively slowly from the top to the bottom of the screen before a rapid flyback to the top of the screen. During the time taken for the Y deflection to go from top to bottom of the screen the X deflection input has moved the spot from left to right across the screen many times. The result of both these inputs is that the spot zig-zags down the screen. During its travel the electron beam is switched on or off by an input to the modulation electrode. The result is that a 'picture' can be painted on the screen.

The above is a description of the basis of a monochrome monitor with what is called a *raster display*. A colour monitor has a screen coated with dots of three different types of phosphor. One type emits red light, one green light and the other blue light. Three electron guns are used, one for each type of phosphor. Dots of the these three types are arranged

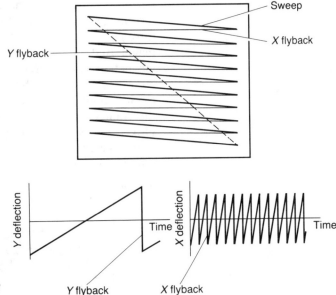

Fig. 7.22 Raster display

in clusters. If the three beams energise all three the appearance is of white light, if just the red phosphor is energised red light.

In order for a stationary picture to remain painted on the screen it is necessary to keep re-energising the phosphor particles. This type of display is produced by what is called a *refresh-type* of cathode-ray tube. An alternative way of retaining the picture on the screen is to use a *storage-type* of cathode-ray tube.

Magnetic-tape recorders

The magnetic-tape recorder can be used to record both analogue and digital signals. It consists of a recording head which responds to the input signal and produces corresponding magnetic patterns on magnetic tape, a replay head to do the converse job and convert the magnetic patterns on the tape to electrical signals, a tape-transport system which moves the magnetic tape in a controlled way under the heads, and signal conditioning elements such as amplifiers and filters.

The recording head consists of a core of ferromagnetic material which has a non-magnetic gap (Fig. 7.23). The proximity of the magnetic tape to the non-magnetic gap means that the magnetic circuit becomes the core and that part of the magnetic tape in the region of the gap. Electrical signals are fed to a coil which is wound round the core and result in the production of magnetic flux in the magnetic circuit, and so that part of the magnetic tape in the region of the gap. The magnetic tape is a flexible plastic base coated with a ferromagnetic powder. When there is magnetic flux passing through a region of the tape it becomes permanently magnetised, i.e. has a remanent flux density. Hence a magnetic record is produced of the electrical input signal. The recording head and the replay heads have similar forms of construction. Thus when a piece of magnetised tape bridges the nonmagnetic gap then magnetic flux is induced in the core. Flux changes in the core induce e.m.f.s in the coil wound round the core. Thus the output from the coil is an electrical signal which is related to the magnetic record on the tape.

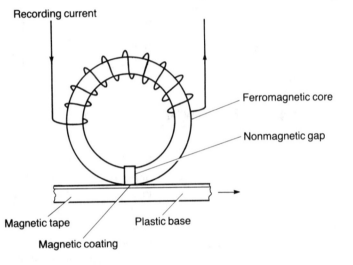

Fig. 7.23 Basis of a magnetic recording head

Figure 7.24(*a*) shows the relationship between the remanent flux density on the magnetic tape and the current in the coil wound round the recording or playback head. (Note: the horizontal axis is more often represented as H, the magnetising force, with H equal to NI where N is the number of turns and I the current.) If the input signal has a current which fluctuates about the zero then the fluctuations in remanent magnetism produced on the tape are not directly proportional to the current. There is, in particular, distortion for very small currents. This can be reduced by adding a steady d.c. current to the signal, i.e. a *d.c. bias*. This shifts the signal to more linear parts of the graph. An alternative to this is to add a high-frequency a.c. current to the signal and give an *a.c. bias*. This also results in the signal being shifted to more linear parts of the graph (Fig. 7.24(*b*)).

If the input signal is sinusoidal with a frequency f then a sinusoidal variation in magnetisation is produced along the tape. The time interval for one cycle is $1/f$ and so if the tape is moving with a constant velocity v then the distance along the tape taken by one cycle is v/f. This distance is called the *recorded wavelength*.

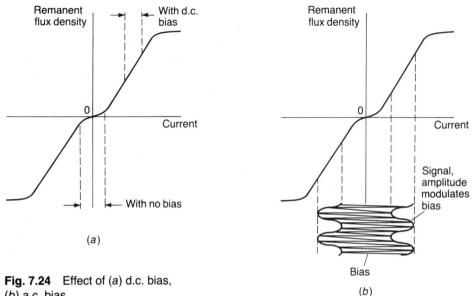

Fig. 7.24 Effect of (a) d.c. bias, (b) a.c. bias

$$\text{Recorded wavelength} = \frac{v}{f}$$

The minimum size this recorded wavelength can have is the gap width. At this wavelength the average magnetic flux across the gap is the average of one cycle and this has a zero value. The size of the gap thus sets an upper limit to the frequency response of the recorder at a particular tape velocity. Typically tape velocities range between about 23–1500 mm/s with a gap width of 5 μm, hence upper frequency limit ranges from about 4.6 kHz up to 300 kHz.

The output from the replay head is proportional to the rate of change of flux ϕ in the head core, i.e. $d\phi/dt$. With a recording of a sinusoidal current input there is a sinusoidal variation of flux on the recording tape. Hence movement of the tape past the recording head will produce a sinusoidal variation of flux in the core. Thus

$$\phi = \phi_m \sin\omega t$$

Head output is proportional to $d\phi/dt$

Head output is thus proportional to $\omega\phi_m \cos\phi t$

where $\omega = 2\pi f$, f being the frequency, and ϕ_m is the maximum flux. Thus the head output depends not only on the flux recorded on the tape but also the frequency. The greater the frequency the greater the output. To overcome this *amplitude equalisation* is used. This is an amplifier with a transfer function which decreases as the frequency increases.

A consequence of the replay head output being proportional to the frequency is that for very low frequencies the output may be very small and so high equalisation amplification is required. This will also amplify noise picked up by the replay head. A consequence of this is that there is a lower frequency limit for which the recorder can be used. This is generally about 100 Hz.

The above is a discussion of what is termed *direct-recording*. With this form the input signal directly determines the magnetic flux recorded on the tape. An alternative to this is *frequency modulation* (see Ch. 4). With this the carrier frequency is varied in accordance with the fluctuations of the input signal. Thus the input signal is modulated before entering the recording head and demodulated after the replay head. Because the carrier frequency is many kilohertz the direct-recording problem of dealing with low frequencies does not occur. Frequency modulation recording can thus be used down to 0 Hz, i.e. d.c. However, the upper frequency limit is less than with direct recording. Typically it is about a third of the carrier frequency, i.e. in the region 2–80 kHz. Frequency modulation tends to give a better signal to noise ratio than with direct recording. Particularly vital with frequency modulation is the control of the tape speed since fluctuations in this can lead to apparent frequency fluctuations.

Recorders generally have more than one recording head. The heads are spaced across the tape and thus each lays down a track of magnetisation. Thus several different signals can be simultaneously recorded.

Digital recording involves the recording of signals as a coded combination of bits, i.e. on or off, high or low signals, 0 or 1. A commonly used method is the *nonreturn-to-zero* (NRZ) method. With this system the flux recorded on the tape is either at the positive saturation value or the negative saturation value (Fig. 7.25). The method uses no change in flux to represent 0 and a change in flux 1. Figure 7.26 illustrates this for the number 0110101. Since the output from the replay head depends on the rate of change of flux on the tape, outputs only occur where the recorded tape has a change of flux. Thus the output is a pulse whenever a 1 was recorded. Digital recording has the advantages over analogue recording of higher accuracy and relative insensitivity to tape speed.

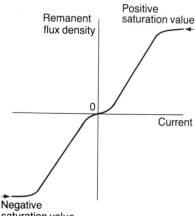

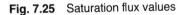

Fig. 7.25 Saturation flux values

Example 7

What is the limiting frequency for a magnetic-tape recorder which has a gap of width 5 μm and is used with a tape speed of 95.5 mm/s?

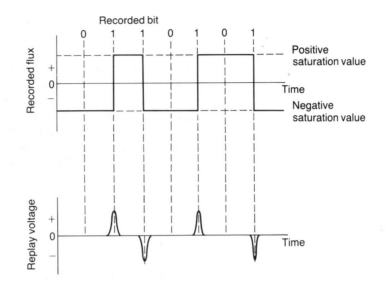

Fig. 7.26 Non-return-to-zero recording

Answer

Using the equation developed above

$$\text{recorded wavelength} = \frac{v}{f}$$

The minimum recorded wavelength is the gap width. Hence the maximum frequency is

$$\text{maximum frequency} = \frac{95.5 \times 10^{-3}}{5 \times 10^{-6}} = 19.1\,\text{kHz}$$

Example 8

A magnetic-tape recorder offers the option of being operated in the direct or frequency-modulated modes. Which mode should be used if the signal to be recorded is (*a*) slow-varying d.c., (*b*) a high-frequency 200 kHz signal?

Answer

(*a*) The direct mode of operation will have a bandwidth which starts at about 100 Hz and so is not suitable for slow varying d.c. The frequency-modulated mode can operate down to d.c. and so is the choice.

(*b*) The frequency-modulated mode has a bandwidth which is unlikely to extend beyond about 80 kHz, while the direct mode goes to much higher frequencies. The direct mode is the choice.

Digital printers

Analogue chart recorders give records in the form of a continuous trace, digital printers give records in the form of numbers, letters or special characters. Such printers are known as *alphanumeric printers*. There are a numbers of versions of such printer, the most commonly used being the *dot-matrix printer*. With such a printer the print head consists of either 9 or 24 pins in a vertical line. Each pin is controlled by an electromagnet which when turned on propels the pin onto an inking ribbon. This impact forces a small blob of ink onto the paper behind the ribbon. A character is formed by moving the print head across the paper and firing the appropriate pins. The 24-pin head uses more dots to form the character and so gives a better-quality image, the individual dots being not discernible to the naked eye. With the 9-pin head this is not the case.

Data loggers

A data logger (Fig. 7.27) consists essentially of a multiplexer, sample and hold element and an analogue-to-digital converter (see Ch. 4 for details of these elements). The unit can monitor a large number of variables. Inputs from the individual transducers, after suitable signal conditioning, are fed into the multiplexer. The multiplexer selects one signal, as a result of some control signal. The sample and hold element then takes a sample of the signal and holds it long enough for the analogue-to-digital conversion to take place without error due to input fluctuations. The output from the unit is thus a digital signal for one of the transducers. The multiplexer can be switched to each transducer in turn and thus digital outputs can be obtained for each. These outputs can be fed to a digital recorder for display and possibly alarm systems. An alternative to this is to feed the outputs to a computer which can then not only store the data but do some processing of it. The output from the computer might then be to a printer and alarm systems or, if it is used as part of a control system, to control elements such as valves.

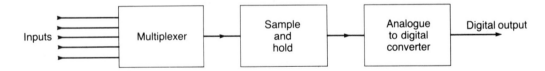

Fig. 7.27 Basic data logger

Typically a data logger may handle 20–100 inputs, though some used with a computer may handle many more. It might have a sample and conversion time of $10 \, \mu s$ and a slew rate of 0.5 V/s. The *slew rate* is the maximum rate of change of input voltage that can be followed. The accuracy is typically about 0.01% of full-scale input and linearity $\pm 0.005\%$ of full-scale input. Cross-talk is typically 0.01% of full-scale input on any one input. The term *cross-talk* is used to describe the interference that can occur when one transducer is sampled as a result of signals from the other transducers.

Problems

1 The following are extracts from the specifications of instruments. Explain the significance for the use of such instruments of the terms and values quoted.

(a) Digital meter
Sample rate 5 readings/second.

(b) Closed-loop servo recorder
Dead band $\pm 0.2\%$ of span.

(c) Data logger
Number of inputs 100
Maximum slew rate 0.4 V/s
Cross-talk on any one input 0.01% of full-scale input.

(d) Magnetic-tape recorder
With direct record/reproduce:

Tape speed mm/s	Signal band Hz
1524	300–300 000
762	200–150 000
381	100–75 000
190.5	100–37 500
95.5	100–18 750
47.6	100–9 300
23.8	100–4 650

(e) Double-beam oscilloscope
Vertical deflection: two identical channels
Bandwidth d.c. to 15 MHz
Deflection factor 10 mV/div to 20 V/div in 11 calibrated steps
Time base: 0.5 µs/div to 0.5 s/div in 19 calibrated steps.

2 List some of the main sources of error that can occur with a moving-coil meter.

3 Compare the characteristics of digital voltmeters which are based on the methods of (a) successive approximations, (b) ramp, (c) voltage to frequency, (d) dual slope.

4 Compare the characteristics of the (a) galvanometric (b) knife-edge and (c) ultraviolet forms of recorders.

5 A galvanometer in a recorder has a resistance of $80\,\Omega$ and a damping factor of 0.7 when the transducer has a resistance of $200\,\Omega$. How is the damping and steady state sensitivity of a recorder galvanometer affected by a resistance of $100\,\Omega$ being included in series with the transducer?

6 The coil of a UV galvanometric recorder has a resistance of $60\,\Omega$ and a damping factor of 2.8 when there is no external circuit connected to it. What additional resistance should be added to give the optimum damping factor of 0.7 when a transducer system of resistance $120\,\Omega$ is connected to the recorder?

7 Explain the significance of the damping factor in the response of

a moving-coil galvanometer to a step input.

8 Explain the significance of the natural frequency of a galvano-meter coil to its response to alternating signals.

9 Explain the effect on the characteristics of a dual-trace oscilloscope of it being alternate mode or chopped mode.

10 Explain what is meant by rise time and overshoot in relation to the Y deflection-plate system of a cathode-ray oscilloscope.

11 What is the significance of the gap width in determining the upper frequency limit of use of a magnetic-tape recorder?

12 What are the advantages and disadvantages of using a magnetic-tape recorder in frequency-modulated mode as compared with the direct mode?

8 Measurement systems

Designing measurement systems

In designing measurement systems it is worth first considering what the input and output from the entire system need to be, then the elements within the system that will fit together and give that input and output. The design procedure thus follows a number of steps.

1 *Identify the nature of the measurement required* For example the variable to be measured, the nominal value, the range of values, the accuracy required, the required speed of measurement, the reliability required, the environmental conditions under which the measurement is to be made.

2 *Identify the required output signal from the system* This means considering what form of display is required and whether the measurement is needed as part of a control system. For example, control applications might require a 4–20 mA current.

3 *Select an appropriate transducer* The transducer needs to match the specification of the input arrived at in step 1 and also be amenable, with suitable signal conditioning, to give the required output arrived at in step 2. Factors that will need to be considered are: range, accuracy, linearity, speed of response, reliability, maintainability, life, power supply requirements, ruggedness, availability, cost.

4 *Select appropriate signal conditioning* This needs to take the output from the transducer and modify the signal in such a way that it delivers the required output.

Temperature-measurement systems

The following are outlines of the characteristics of commonly used industrial temperature-measurement systems.

Bimetallic strips

This device consists of two different metal strips bonded together. The metals have different coefficients of expansion and so when the temperature changes the composite strip deforms. This deformation may be used as a temperature controlled switch, as in the simple thermostat used with many domestic central-heating systems (see Fig. 5.5). The composite strip may be in the form of a spiral so that the unwinding or winding up of the spiral when the temperature changes can be used to directly move a pointer across a scale (Fig. 8.1). Bimetallic strip devices are robust, can be used within a range of about − 30 °C to 600 °C, can be used for two-step control, have an accuracy of the order of ± 1%, are fairly slow reacting to change, and are relatively cheap. They have an advantage over mercury in glass or mercury in metal thermometers that a breakage does not result in mercury leaking and so the possibility of poisonous fumes.

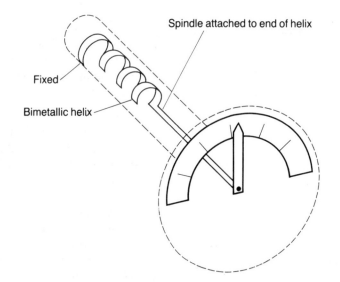

Fig. 8.1 Bimetallic thermometer

Liquid in glass or metal

The liquid in glass thermometer is direct reading, fragile, capable of reasonable accuracy under standardised conditions, fairly slow reacting to change and cheap. With mercury as the liquid the range is − 35 °C to 600 °C, with alcohol − 80 °C to 70 °C, with pentane − 200 °C to 30 °C.

Liquid in metal thermometers are more robust, the thermometer bulb being generally connected to a Bourdon tube (see Fig. 3.20(e)) and filled with the appropriate liquid. When the liquid expands the Bourdon tube straightens out to some extent and can cause a pointer to move across a scale. With mercury the range is − 39 °C to 650 °C, with alcohol − 46 °C to 150 °C. Accuracy is about ± 1% of full-scale reading.

Gas in metal

The industrial form of a gas thermometer is usually a thermometer bulb connected to a Bourdon gauge (see Fig. 3.20(e)) and filled with gas, e.g. nitrogen. When the temperature rises the gas pressure increases and this causes the Bourdon tube to straighten out to some extent and can cause a pointer to move across a scale. The bulb of the thermometer is fairly large, about 50–100 cm^3. The thermometer is robust, has a range of about $-100\,°C$ to $650\,°C$, is direct reading, can be used to give a display at a distance, and has an accuracy of about $\pm 0.5\%$ of full-scale deflection. When used at a distance a mechanism has to be used to correct for the temperature gradient along the length of tube connecting the thermometer bulb to the Bourdon tube. A common way of doing this is by the use of two Bourdon tubes. One of the Bourdon tubes is connected to the thermometer bulb while the other is connected to just a tube which runs parallel with the connecting tube to the thermometer bulb. The instrument display is then driven by the difference in movement between the two Bourdon tubes.

Vapour pressure

The vapour pressure consists of a thermometer bulb connected to a Bourdon gauge (see Fig. 3.20(e)) and containing a small amount of liquid. The higher the temperature the greater the amount of liquid that has evaporated and the greater the pressure exerted by its vapour. A higher pressure causes the Bourdon tube to straighten out to some extent and can cause a pointer to move across a scale. With methyl chloride as the liquid the range is about 0–50 °C, with ethyl alcohol 90–170 °C, toluene 150–250 °C. The scale is non-linear and there is an accuracy of about $\pm 1\%$. The instrument is robust, direct reading and can be used at a distance.

Resistance

Resistance thermometers (see Ch. 3 and Fig. 3.1) using metal wire coils are generally used with a Wheatstone bridge signal conditioner (see Ch. 4). Compensation is generally necessary for the temperature gradient along the leads to the thermometer coil, Fig. 4.5 shows one method. The range and sensitivity of the system depends on the metal used. With nickel the range is $-80\,°C$ to $320\,°C$, copper $-200\,°C$ to $260\,°C$, platinum $-200\,°C$ to $850\,°C$. Non-linearity is generally less than about 1% of the full-scale range. Accuracy is typically about $\pm 0.75\%$.

Resistance thermometers using thermistors (see Ch. 3 and Figs 3.2 and 3.3) give much larger resistance changes per degree than metal wire elements. However the variation of resistance with temperature is markedly non-linear. Wheatstone bridges (see Ch. 4) are commonly used as the signal conditioner. Their small size means a small thermal capacity and hence a rapid response to temperature changes. The temperature range over which they can be used will depend on the thermistor concerned, ranges within about $-250\,°C$ to $650\,°C$ are possible. There is a tendency for the calibration to change with time.

Thermoelectric

Thermocouples (see Ch. 3 and Figs 3.17 and 3.18) are generally used with an electronic circuit or potentiometer circuit as a signal conditioner. Because the temperature indicated is that at the junction between two dissimilar metals, there is a very small thermal capacity and so a rapid response to temperature changes. Depending on the materials used ranges within about $-200\,°C$ to $2200\,°C$ are possible. For example, chromel–constantan has a range of 0–$980\,°C$ and a sensitivity of $63\,\mu V/°C$, iron–constantan a range $-180\,°C$ to $760\,°C$ and a sensitivity of $53\,\mu V/°C$, platinum–platinum/ rhodium 13% a range 0–$1759\,°C$ and a sensitivity of $8\,\mu V/°C$. Base metal thermocouples are relatively cheap but deteriorate with age. They have accuracies of the order of ± 1–3%. Noble metal thermocouples are more expensive, more accurate (± 1% or less), and more stable. Thermocouples have the disadvantage of the need for a cold junction with a temperature that is controlled or for some means of compensating for changes in cold junction temperature. Often the cold junction is the ambient temperature which is then measured with a thermometer and a correction factor applied to take account of the difference of this temperature from $0\,°C$ (see Ch. 3 for a discussion of this).

Radiation

Objects emit electromagnetic radiation, the total rate of emission per second being related to the temperature of the object (Fig. 8.2). With the *radiation pyrometer* the radiation from the object is focused onto a radiation detector (Fig. 8.3). The radiation detector might be a resistance thermometer, a thermistor, or a thermopile (this is a group of thermocouples in series). The thermal detectors give a *broad band detector* since the radiation is detected over a wide band of frequencies. The output from the detector is thus the summation of the power emitted at every wavelength and hence is represented by the area under the Fig. 8.2 graph at a particular temperature. This is proportional to the fourth power of the temperature, the temperature being on the kelvin scale. Thus the output from the detector is a measure of the temperature of the object.

An alternative to these forms of thermal detector is what is termed a photon detector, e.g. photoconductive and photo-emissive cells. These rely on atomic processes for their detection of the radiation and so are much quicker at respond-

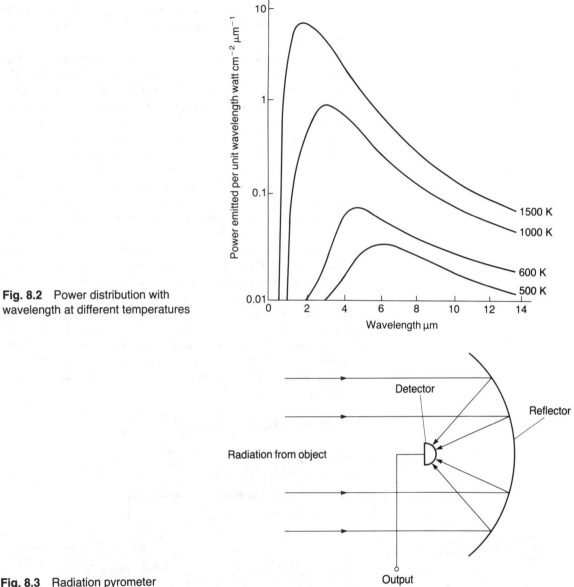

Fig. 8.2 Power distribution with wavelength at different temperatures

Fig. 8.3 Radiation pyrometer

ing than the thermal detectors which rely on temperature measurement. The difference may be several microseconds compared with several milliseconds. The photon detectors give a *narrow band detector* since they are responsive to only a narrow band of wavelengths, a filter sometimes being added to narrow the band yet more. The output of such a detector is proportional to

$$\frac{\Lambda\lambda}{\lambda^5} \exp(-\ C/\lambda T)$$

where λ is the centre of the wavelength pass band, $\Delta\lambda$ the width of the band, C a constant and T the temperature on the kelvin scale.

In some forms of instrument a rotary mechanical disc or shutter is used to chop the radiation before it reaches the detector. The result is that the output from the detector is alternating. This form of output is more readily amplified than the d.c. output that otherwise would be produced. The accuracy of a radiation pyrometer is typically of the order of $\pm 0.5\%$ and ranges are available within the region 0–2000 °C.

The output of the above pyrometers is affected by changes in the emissivity of the hot object and changes in the transmission characteristics of the intervening medium between object and pyrometer. For this reason calibration is necessary for a particular situation. A system that overcomes these problems is the *two-colour pyrometer* (Fig. 8.4). The incoming radiation from the object is split into two equal parts. One band is then transmitted through a narrow band filter which allows only the radiation of a certain wavelength through. The other band also passes through a narrow band filter, but one which allows a different wavelength through. The outputs from both filters are fed to detectors which respond to those wavelengths, possibly thermal or photon detectors. The ratio of the two outputs is then used as a measure of the temperature of the object.

$$\text{Output ratio} = \left(\frac{\lambda_2}{\lambda_1}\right)^5 \exp \frac{C}{T}\left(\frac{1}{\lambda_2} - \frac{1}{\lambda_1}\right)$$

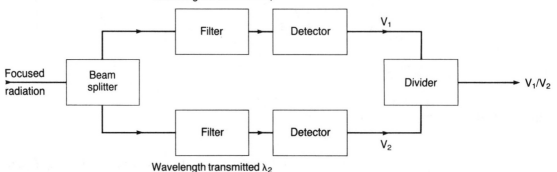

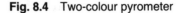

Fig. 8.4 Two-colour pyrometer

where λ_1 and λ_2 are the centre wavelengths of the two filters, T is the temperature on the kelvin scale and C is a constant with the value $14\,388\,\mu\text{m K}$.

Example 1

Suggest a measurement system that could be used to give a measurement of the temperature in an oven. The oven temperature range is room temperature to 200 °C and the temperature is to be displayed on a dial on the front of the oven. The oven temperature does not change rapidly. The accuracy required is ± 2 °C. The system needs to be robust and cheap.

Answer

The nature of the measurement is: room temperature to 200 °C with an accuracy of ± 2 °C, i.e. ± 1% of full-scale reading, and a slow rate of temperature change. The output required is a display on a dial. Possible transducers, bearing in mind the need for robustness, are bimetallic strips, liquid in metal, gas in metal, vapour pressure, and resistance. Liquid in metal, gas in metal and vapour pressure would have the disadvantage of possibly contaminating the interior of the oven in the event of a leakage. They do however have the advantage of allowing the transducer to be some distance from the display. The resistance transducer would require a d.c. supply. The vapour pressure and a resistance element, if a thermistor, would give a non-linear scale. In considering signal conditioning and the requirement for a cheap system, an obvious contender is a bimetallic strip in the form of a helix. This can directly drive a display pointer. The liquid or gas in steel transducer could use a Bourdon tube to translate the change in pressure into displacement to enable a pointer to move across a scale. Bearing in mind these points it is likely that the optimum choice could be liquid or gas in metal.

Example 2

As part of an alarm system for overheating of liquid in a container, a measurement system is required which will set off an alarm when the temperature rises above 40 °C. The liquid is normally at 30 °C. The output from the measurement system must be a 1 V signal. Suggest a possible measurement system.

Answer

The nature of the measurement is a measurement of the temperature of a liquid, temperature change from 30 °C to 40 °C to be detected and since it is for an alarm system a reasonable speed of response is presumably required. The output signal is to be a 1 V signal when the temperature reaches 30 °C. Since the output is to be electrical an obvious possibility for a transducer is an electrical resistance element. To generate a voltage output the resistance element could be used with a d.c. Wheatstone bridge and the out-of-balance voltage used. This will probably be less than 1 V for a temperature change from 30 °C to 40 °C and so amplification will be required. This could be an operational amplifier.

Suppose a high sensitivity nickel element is used. Nickel has a temperature coefficient of resistance of 0.0067/°C. Thus if the resistance element is taken as being 100 Ω then the change in resistance in going from 30 °C to 40 °C will be about 6.7 Ω. If this element forms one arm of a Wheatstone bridge, then the out-of-balance voltage is given by (see Ch. 4):

$$\delta V_{\mathrm{o}} = \frac{V_{s} \delta R_{1}}{R_{1} + R_{2}}$$

If all the arms of the bridge have the same nominal resistance at 30 °C and so the bridge is balanced at that temperature, then with a d.c. supply of 4 V the out-of-balance voltage is 0.109 V. To amplify this tol V requires an amplifier with a transfer function of $1/0.109 = 9.17$. A differential amplifier, as in Fig. 4.15, could be used. To give the required transfer function it could have a feedback resistance of 9.17 kΩ and input resistances of 1 kΩ .

Example 3

Suggest a possible measurement system that could be used to give a display on a scale of the temperature of molten iron in a furnace. The scale is to be located some distance from the furnace.

Answer

The nature of the measurement is a measurement of the temperature of molten iron in a furnace. This is likely to be at temperatures of more than 1100 °C and probably in excess of 1500 °C. The environment is hostile. The output is to be a reading on a scale some distance from where the measurement is made. The obvious possibility is a radiation pyrometer. This could be broad band, narrow band or two colour. If a fast response was required then narrow band or two colour would be better. If there was to be no calibration then the two-colour pyrometer is indicated. Signal conditioning would depend on the transducer used and the distance over which the signals are to be transmitted. Thus, if the transducer used was a photoconductive cell the resulting change in resistance can be transformed into a voltage as the out-of-balance voltage of a Wheatstone bridge. This voltage can then be amplified by a differential amplifier (as in the previous example) before being transmitted to the meter.

Example 4

A proposed measurement system consists of a thermocouple, an amplifier and a chart recorder. What will be the accuracy of the resulting measurement if the thermocouple has a non-linearity error of ±1% of full-scale reading, the amplifier and chart recorder combined ±1% of full-scale reading? The full-scale reading of the arrangement is set to be 0–100 °C.

Answer

As indicated in Chapter 1, the accuracy of the system will be the sum of the accuracies of the constituent parts. Thus the accuracy will be ±2%, i.e. ±2 °C.

Pressure measurement

The following are methods commonly used to measure pressures which are close to or greater than atmospheric pressure. The term *absolute pressure* is used for the pressure measured relative to zero pressure, the term *gauge pressure* for the pressure measured relative to atmospheric pressure. At the surface of the earth the atmospheric pressure is generally about 100 kPa. This is sometimes referred to as a pressure of 1 bar.

Manometers

Figure 8.5 shows two forms of manometer, the simple *U-tube manometer* and an industrial form of it. With the simple U-tube manometer the pressure difference between the two limbs is indicated by a difference in vertical level h of the manometric liquid in the two limbs.

$$\text{Pressure difference} = h\rho g$$

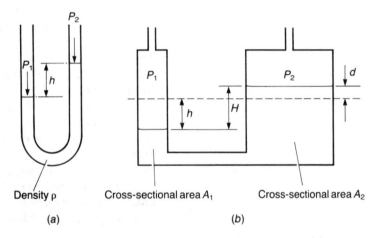

Fig. 8.5 (a) U-tube manometer, (b) industrial manometer

where ρ is the density of the manometric liquid and g the acceleration due to gravity. If one of the limbs is open to the atmosphere then the pressure difference is the gauge pressure. With the industrial manometer the arrangement is still a form of U-tube but one of the limbs has a much greater cross-sectional area than the other. For such an arrangement

$$\text{pressure difference} = H\rho g = (h + d)\rho g$$

where d is the amount by which the level of the liquid in the larger cross-section tube changes and h that movement for the smaller cross-section tube, from when the pressures in the two limbs were the same. Since the volume of liquid leaving one limb must equal the volume entering the other,

$$A_1 h = A_2 d$$

where A_1 and A_2 are the cross-sectional areas of the two limbs.

Hence

$$\text{pressure difference} = [(A_2 d/A_1) + d]\rho g$$
$$= [(A_2/A_1) + 1]d\rho g$$

Thus the movement of the liquid level d in the wide tube from its initial zero level is proportional to the pressure difference. This displacement can be indicated by means of a float-and-lever system and thus move a pointer across a scale.

Manometers, with an appropriate manometric liquid, can be used to measure pressure differences of the order of 20 Pa to 140 kPa. Water, alcohol, or mercury are commonly used as the manometric liquid. Corrections need to be made for the variation of density of the manometric liquid with temperature, its thermal expansion and evaporation, the effect of altitude on the value of the acceleration due to gravity, nonverticality of the tubes and measuring scales, and additionally there are difficulties in taking accurate readings due to the meniscus of the manometric liquid. Accuracy is typically about ±1%.

Bourdon tubes

The Bourdon tube (see Ch. 3 and Fig. 3.20(*e*)) is a very widely used transducer for the measurement of pressure. The Bourdon tube may be in the form of a C, a flat spiral, a helical spiral or twisted. Whatever the form, an increase in the pressure inside the tube results in the tube straightening out to an extent which depends on the pressure change. The displacement of the end of the tube may be used to directly move a pointer across a scale or to move the slider of a potentiometer or to move the core iron rod in a linear variable differential transformer (see Ch. 3 and Fig. 8.6). These last two will give an electrical output related to the pressure.

Fig. 8.6 Bourdon tube with (*a*) a potentiometer, (*b*) linear variable differential transformer

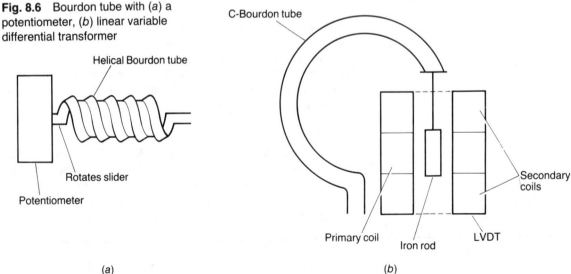

(*a*)

(*b*)

Typically, Bourdon tube measurement systems are used for pressure differences in the range $10^4–10^8$ Pa. They are robust with an accuracy of about $\pm$ 1% of full-scale reading.

Diaphragms, capsules and bellows

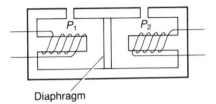

Fig. 8.7 Variable reluctance differential pressure sensor

See Chapter 3 for a discussion of these forms of transducers. Figure 8.7 shows one form of a diaphragm pressure sensor. The diaphragm responds to the pressure difference between its two sides by flexing and the centre portion becomes more displaced to one side than the other. In the figure this displacement gives rise to a change in reluctance, increasing that on one side of the diaphragm and decreasing it on the other. Other forms of this differential pressure sensor can result in capacitance changes or, if strain gauges are attached to the diaphragm, changes in resistance (see Fig. 3.6). With reluctance or capacitance changes the signal conditioner used is likely to be an a.c. bridge (see Ch. 4) with its out-of-balance signal being amplified. With the strain gauges the signal conditioner is likely to be a Wheatstone bridge involving four active gauges (as in Fig. 4.7), the out-of-balance signal being amplified. Diaphragm instruments in general are used in the range $1–10^8$ Pa and have an accuracy of about $\pm 0.1\%$. Capacitive versions tend to have ranges of about 1 Pa to 200 kPa, reluctance versions $1–10^8$ Pa and strain gauge versions $10^4–10^8$ Pa. Capacitive, reluctance and strain gauge versions can be used with frequencies up to about 1 kHz.

Semiconductor strain gauges, based on the piezo-resistive effect, can be used to detect the movement of a diaphragm. While such gauges could be cemented to the surface of the diaphragm, it is now more customary to use a silicon sheet as the diaphragm and introduce doping material into the silicon, at appropriate places, and so produce the strain gauges integral with the diaphragm. Such gauges can be used with a four-active-gauge Wheatstone bridge. Typically such a gauge can respond to pressure differences up to about 10^5 Pa with an accuracy of about $\pm 0.5\%$. With the Wheatstone bridge the typical out-of-balance voltage is a few millivolts for each kilopascal pressure difference. Response time is about 0.1 ms.

Another form of pressure gauge based on the use of a diaphragm is the *closed-loop differential pressure cell*. This is also referred to as a force-balance system or torque-balance system. Figure 8.8 shows one version of such a cell. Movement of the diaphragm causes the force beam to deflect about its pivot. This movement results in a change in reluctance. This then results in an out-of-balance signal from an a.c bridge. This signal takes the form of an alternating signal whose amplitude is related to the out-of-balance condition. After amplification the signal is then demodulated. This means removing the alternating signal supplied by the source for the a.c bridge. The result is then fed to a display. The signal is also fed to a solenoid which moves a pivoted beam, the feedback beam. This in turn acts on the force beam. The result is that the force beam is acted on by torques due to the difference in pressures across the diaphragm and the opposing torque produced by the feedback beam. At equilibrium these torques balance. When this occurs the current is related to the difference in pressure by the relationship

$$I = K(P_1 - P_2) + C$$

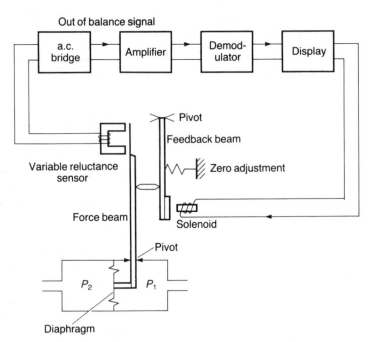

Fig. 8.8 Closed-loop differential pressure cell

where K and C are constants. Other versions of the cell are available. There is, for example, a pneumatic version. In place of the reluctance sensor for the force beam displacement a flapper-nozzle arrangement is used and the feedback to the feedback beam is by pressure in a bellows. For such an arrangement the output pneumatic pressure p is related to the difference in pressure being measured by

$$p = K(P_1 - P_2) + C$$

Typically, such cells have ranges within 0–10^5 Pa with accuracies of about $\pm 0.2\%$ and response times of the order of a second.

Piezo-electric crystals produce a potential difference across opposite faces when squeezed. If the movement of a diaphragm results in such a crystal being squeezed then the output is a potential difference related to the flexure of the diaphragm which in turn is related to the pressure difference between its two faces. A typical instrument based on this principle will respond to pressures up to about 100 MPa and can handle fluctuating pressures changes within a bandwidth of 5 Hz to 500 kHz.

A capsule can be considered to be just a more sensitive version of the diaphragm with a bellows being just a stack of capsules and so even more sensitive. Figure 8.9 shows one form of bellows pressure sensor. Pressure changes in the bellows cause the sealed end of the bellows to become displaced. This in turn moves the core rod in a linear variable differential transformer (see Ch. 3) and so gives an output which is related to the pressure in the bellows. Another form uses the movement of the end of the bellows to move the slider of a potentiometer. Bellows instruments are used to measure pressure differences in the range 200–10^6 Pa with an accuracy of about ± 0.1%. They have poor zero stability.

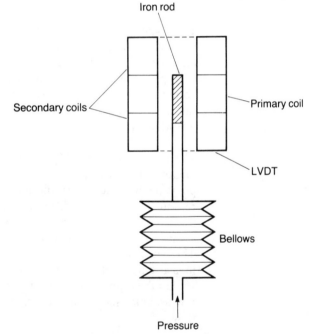

Fig. 8.9 Bellows pressure sensor

Example 5

Suggest a sensor which would be suitable for the measurement of pressures which are fluctuating at a frequency of the order of 100 kHz.

Answer

The piezo-electric form of diaphragm pressure sensor is capable of responding to pressure fluctuations at this frequency.

Example 6

Suggest a pressure measurement system that could be used to measure pressure differences in the range 5–10 kPa and provide an electrical output which could be used with a chart recorder. The pressure changes with time are no more than 10 Hz.

Answer

The measurement is that of pressure in a range 5–10 kPa and which changes at less than 10 Hz. The output needs to be electrical and capable of operating a chart recorder. A possible transducer is a diaphragm with some form of sensor for the diaphragm movement. The signal conditioner will depend on the sensor selected. For example, with the strain gauge version a Wheatstone bridge would be used with a differential amplifier. A typical commercial system would have ranges of 2–70 kPa and 100–1000 Pa and give a full-scale output of 40 mV when the Wheatstone bridge had a d.c. supply of 10 V. Accuracy would be ±0.5%.

Flow measurement

Two types of measurement are grouped under this heading and different techniques are required for each. These are the measurement of the velocity at a point in a fluid by, for example, the Pitot static tube and the measurement of the volume flowrate by, for example, differential flowmeters and turbine flowmeters. See Chapter 3 for details of these transducers.

Pitot static tube

The *Pitot static tube* (Fig. 3.27) is used for velocity measurements, the pressure difference being measured between a point in the fluid where the fluid is in full flow and a point at rest in the fluid. The difference in pressure is proportional to the square of the velocity for an incompressible fluid, for a compressible fluid (a gas) the relationship needs modification. The electrical techniques used with diaphragms for the measurement of differential pressure are often used. Pitot static tubes can be used for measurements of fluid velocities as low as 1 m/s and as high as 60 m/s and for both liquid and gas flow.

Differential pressure

There are a number of forms of flowmeter based on the measurement of the pressure difference between the flow in a full cross-section tube and that at a constriction, the *Venturi tube* (Fig. 3.23), the *orifice flowmeter* (Fig. 3.24), the *nozzle flowmeter* (Fig. 3.25) and the *Dall flowmeter* (Fig. 3.25). The Venturi tube offers the least resistance to fluid flow and thus has the least effect on the rate of flow of fluid through the pipe while the orifice plate offers the greatest resistance. A typical flowmeter system consists of one of the above with a differential pressure sensing element, e.g. a diaphragm with a sensor to detect its movement. All the flowmeters have long life without maintenance or recalibration being required and an accuracy of about ±0.5%.

The *rotameter* (Fig. 3.26) is another form of constriction flowmeter and employs a float in a tapered vertical tube. The float moves up the tube to a height which depends on the rate of flow. A scale alongside the tube can thus be calibrated to read directly the flow rate corresponding to a particular height of float. The rotameter can be used to measure flow rates from about 30 ml/s (30×10^{-6} m³/s) to 120 l/s (120×10^{-3} m³/s). It is a relatively cheap instrument, capable of a long life with little maintenance or recalibration being required. It is not highly accurate, about ±1%, and has a significant effect on the flow.

Turbine flowmeter

The turbine flowmeter (Fig. 3.28) is often used with a magnetic pick-up which produces an induced e.m.f. pulse every time a rotor blade passes it. The pulses are counted and so the number of revolutions of the rotor determined. The meter is used with liquids and offers some resistance to fluid flow. It is expensive. The accuracy is typically about ±0.3%.

Positive displacement meters

This form of flowmeter works on the principle of dividing up the flowing fluid into known volume packets and then counting them to give the total volume passing through the meter. They are widely used for water meters, gas meters and petrol pump meters to determine the volume delivered. If the volume delivered over a particular time is monitored then the volume rate of flow can be established. There are many forms of positive displacement meter, e.g. rotating lobe meter, rotating vane meter, nutating disc meter, reciprocating piston meter. Some are used with liquids, others with gases. Figure 8.10 shows the rotating lobe meter, this being used with gases. Accuracies are generally of the order of ±0.2%.

Trapped volume

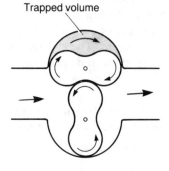

Fig. 8.10   Rotating lobe meter

Example 7

Suggest a flowmeter that could be inserted in a pipe to give a quick, rough, visual indication of the volume rate of flow and that is cheap.

Answer

The obvious choice is a rotameter since the flowrate can be read directly from the position of the float against a scale. Also it is cheap.

Example 8

Suggest a flow measurement system that could be used as part of a control system and would give a pneumatic output related to the volume rate of flow of water through a pipe.

Answer

A differential flowmeter, e.g. a Venturi, with a differential pressure sensor which converts the pressure difference to a pneumatic signal. A possibility is a pneumatic closed loop differential pressure cell (see earlier this chapter).

Example 9

Suggest a system that could be used to determine the volume of liquid delivered through a pipe.

Answer

Some form of positive displacement meter is required. A reciprocating piston, rotating disc or rotating vane is suitable for liquids. The pressure exerted by the fluid can drive the rotating element and this in turn could, through gearing, drive a simple mechanical counter.

Measurement of liquid level

There are a wide variety of methods that can be used to measure the level of a liquid in a container.

Dipsticks

The simple dipstick is just a metal bar with a scale marked on its side. To determine the level of liquid in a container the dipstick is just held vertically in the liquid from a fixed position. The stick is then removed and the mark left by the liquid on the stick enables the position of the liquid level to be determined. An obvious example of the use of a dipstick is the determination of the oil level in a car engine. The dipstick is very cheap and fairly accurate, but does not provide a continuous reading.

Float systems

Figure 8.11 shows one version of a float system for the measurement of liquid level. Changes in level cause the float

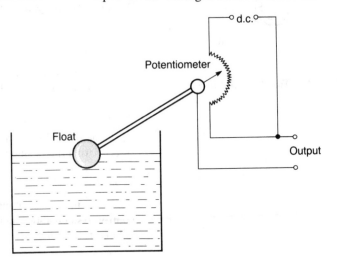

Fig. 8.11 Float system

to move and so move the potentiometer slider over the resistance track. The result is an output voltage which is proportional to the movement of the liquid surface. Such a system is widely used for the determination of the level, and hence amount, of petrol in motor car fuel tanks. The method is fairly accurate.

Figure 8.12 shows two forms of level measurement based on the measurement of differential pressure. In Fig. 8.12(*a*) the differential pressure cell determines the pressure difference between the liquid at the base of the tank and atmospheric pressure when the tank is open to the atmosphere. The pressure difference is $h\rho g$, where h is the height of the liquid above the base of the tank. With a closed, or open, tank the system illustrated in Fig. 8.12(*b*) can be used. Here the pressure difference between the liquid at the base of the tank and the air or gas above the surface of the liquid is measured. This equals $h\rho g$, where h is the height of the liquid surface above the tank base.

A different method is the *bubbler method*. This uses a pipe which dips to virtually the bottom of the tank (Fig. 8.13). A constant flow of air, or some other suitable gas, passes into the tube and bubbles out of the bottom. This escaping gas limits the air pressure in the tube. The air pressure at which this situation is reached is measured using a differential pressure cell, the pressure being proportional to the height of the liquid above the bottom of the tube.

Differential pressure methods

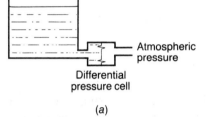

Atmospheric pressure

Differential pressure cell

(*a*)

Closed vessel

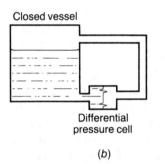

Differential pressure cell

(*b*)

Fig. 8.12 Level measurement using a differential pressure cell

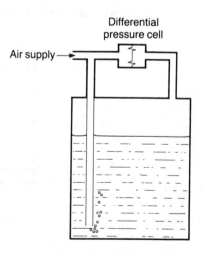

Differential pressure cell

Air supply →

Fig. 8.13 Bubbler method

Capacitive methods

Figure 8.14 shows one form of a capacitive method for the determination of liquid level. With the concentric cylinder capacitor the capacitance changes when the liquid level changes (see Ch. 3). The capacitor can be incorporated in an

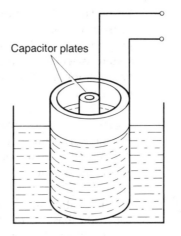

Fig. 8.14 Concentric cylinder capacitor

a.c. bridge and changes in liquid level then result in changes in the amplitude of the out-of-balance output alternating voltage.

Load cell methods

Figure 8.15 illustrates the principle involved in the use of load cells (see Ch. 3) to determine liquid level in a container. Essentially the load cells, which are included in the supports for the container, determine the container weight. Since the weight depends on the level of liquid in the container then the load cells give responses related to liquid level. The load cells deform under the action of the liquid weight and the deformation can be monitored by strain gauges attached to the walls of the load cell. The gauges can be incorporated in a four-active-gauge Wheatstone bridge (see Ch. 4) and the out-of-balance voltage then becomes a measure of the liquid level.

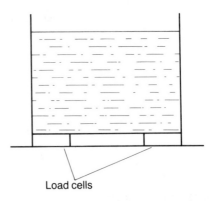

Fig. 8.15 Load cell system

Load cells

Ultrasonic level gauge

In one version of an ultrasonic level gauge an ultrasonic transmitter is placed above the surface of the liquid and emits pulses of ultrasonics (Fig. 8.16). The pulses are reflected from the surface to a receiver. The time taken from emission to reception of the reflected pulse can be measured. Since the

Ultrasonic transmitter and receiver

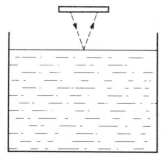

Fig. 8.16 Ultrasonic level gauge

time taken depends on the distance of the liquid surface from the transmitter/receiver the level of the liquid can be determined.

Example 10

Suggest a level-measurement system that could be used as part of a level-control system for water in a tank. The output from the measurement system is to be an electrical signal. The water level fluctuates between a depth of 1.5 m and 2.0 m.

Answer

The nature of the required measurement is a water level varying between 1.5 m and 2.0 m. The required output is an electrical signal. Possible methods could be a float method with a potentiometer, a differential pressure method with a differential pressure cell, a load cell with a Wheatstone bridge and amplifier, and a capacitive method with an a.c. bridge, amplifier and demodulator. Considering the differential pressure method, the gauge pressure ($P = h\rho g$) changes from $1.5 \times 1000 \times 9.8$ to $2.0 \times 1000 \times 9.8$, i.e. 14.7–19.6 kPa. This is well within the range of a differential pressure cell (see earlier this chapter).

Example 11

Suggest a system that could be used to switch an electrical alarm on when the level of water in a tank reaches a particular level. The system needs to be robust and cheap.

Answer

Float systems are particularly robust. Figure 8.17 shows a possible system involving a float. The movement of the arm can be used to activate an electrical switch in the alarm circuit, so switching on the alarm when the level reaches the required level and switching it off when the level falls below that value.

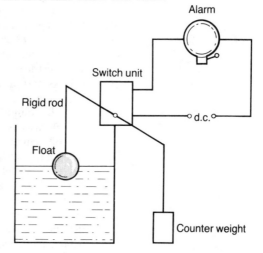

Fig. 8.17 Example 10

Example 12

Suggest a measurement system that could be used to determine the level of a highly corrosive liquid in a large tank.

Answer

The nature of the measurement is level of corrosive liquid in a large tank. Because of the corrosive nature of the liquid it is desirable that the transducer does not have to come into contact with the liquid. One possibility is to use load cells in the supports of the tank and so determine the level by effectively measuring the weight of the tank and contents. The load cell could employ four active strain gauges to form a four-active-gauge Wheatstone bridge. The out-of-balance voltage is then a measure of the level. This might be fed directly to a meter or via an amplifier.

Problems

1   Suggest measurement systems that could be used for the following:
    (a) A low-cost system is required for the measurement of the temperature in a domestic refrigerator and presentation of the result as a pointer on a linear scale. The scale is to be located in the fascia of the refrigerator and the temperature measured half-way up one side of the inside of the refrigerator.
    (b) A low-cost, robust, instrument is required which can be attached to the head of gas cylinders and enable the pressure of gas in the cylinder to be read off a scale. No great accuracy is required.
    (c) As part of a system to control the thickness of rolled sheet, a measurement system is required to determine the thickness of sheet metal as it emerges from rollers. The sheet metal is in continuous motion and the measurement needs to be made quickly so corrective action can be made quickly. The measurement system has to supply an electrical signal which can be fed to a differential amplifier for the control system.
    (d) As part of a control system a system is required to determine the difference in levels between liquids in two containers. The output is to be shown on a meter and provide an electrical signal for the control system
    (e) As part of the control system for a furnace it is necessary to monitor the rate at which the heating oil flows along a pipe. The output from the measurement system is to be an electrical signal which can be used to adjust the speed of the oil pump. The system must be capable of operating continuously and automatically, without adjustment, for long periods of time.
2   A cylindrical load cell, of the form shown in Fig. 3.7, has four strain gauges attached to its surface. Two of the gauges are in the circumferential direction and two in the longitudinal axis direction. When the cylinder is subject to a compressive load, the axial gauges will be in compression while the circumferential

ones will be in tension. If the material of the cylinder has a cross-sectional area A and an elastic modulus E, then a force F acting on the cylinder will give a strain acting on the axial gauges of $-F/AE$ and on the circumferential gauges of $+vF/AE$, where v is Poisson's ratio for the material.

Design a complete measurement system, using load cells, which could be used to monitor the mass of water in a tank. The tank itself has a mass of 20 kg and the water when at the required level 40 kg. The mass is to be monitored to an accuracy of ± 0.5 kg. The strain gauges have a gauge factor of 2.1 and are all the same resistance of 120.0 Ω. For all other items, specify what your design requires.

3   Design a complete measurement system involving the use of a thermocouple to determine the temperature of the water in a boiler and give a visual indication on a meter. The temperature will be in the range 0–100 °C and is required to an accuracy of $\pm 1\%$ of full-scale reading. Specify the materials to be used for the thermocouple and all other items necessary. In advocating your design you must consider the problems of cold junction and non-linearity.

You will probably need to consult thermocouple tables. The following data is taken from such tables and may be used as a guide.

Materials	e.m.f. in mV at					
	0 °C	20 °C	40 °C	60 °C	80 °C	100 °C
Copper–constantan	0	0.789	1.611	2.467	3.357	4.277
Chromel–constantan	0	1.192	2.419	3.683	4.983	6.317
Iron–constantan	0	1.019	2.058	3.115	4.186	5.268
Chromel–alumel	0	0.798	1.611	2.436	3.266	4.095
Platinum–10%Rh,Pt	0	0.113	0.235	0.365	0.502	0.645

4   A suggested design for the measurement of liquid level involves a float which in its vertical motion bends a cantilever. The degree of bending of the cantilever is then taken as a measure of the liquid level. Strain gauges are used to monitor the bending of the cantilever with two gauges being attached to its upper surface and two to the lower surface. The gauges are then to be incorporated in a four-gauge active Wheatstone bridge with the out-of-balance potential difference monitored.

(a) Carry out preliminary design calculations to determine whether the idea is feasible and what values and form the various components may need to take.

When a force F is applied to the free end of a cantilever of length L, the strain on its surface a distance x from the clamped end is

$$\text{strain} = \frac{6(L - x)}{wt^2 E}$$

where w is the width of the cantilever, t its thickness and E

the tensile modulus of the material.

(b) On the basis of your design construct a prototype, test and calibrate it.

5 (a) Design a measurement system which could be used to monitor continuously the temperature in a room over a 24-hour period. The required accuracy is ±0.5 °C and the output from the system should be to a chart recorder.

(b) On the basis of your design assemble a prototype, test and calibrate it.

9 Control systems

Designing control systems

The designing of a control system is a very complex problem. In this chapter the aim is to give an understanding of the principles involved so that design procedures and the behaviour of control systems can be appreciated.

Design can be considered to involve a number of steps. The early steps are generally concerned with a consideration of what is to be controlled and what hardware elements will be needed. Because the feedback transducer and the final control element are connected to and directly interact with the process being controlled, they need to be selected first. The other elements in the control system can then be selected to fit with them.

1 Identify those variables of the process which are to be controlled and the degree of precision required.
2 Identify those properties of the process which are going to be manipulated by the control final elements in order to keep the controlled variables within the required limits.
3 Select the appropriate measurement system for each variable being controlled. See Chapter 8 for the steps that need to be considered for this selection.
4 Select the final control elements needed to provide the required manipulation. Factors that need to be considered are ruggedness, reliability, maintainability, life, mounting and coupling requirements, power supply requirements, input signal characteristics, availability, cost.
5 Select the remaining elements in the control system, taking into account such factors as whether the connections should be by means of electrical, hydraulic or pneumatic signals. Electrical control systems have the advantage that the control mode used can be easily altered, control signals can be transmitted over large distances, an electrical supply is generally readily available but the

disadvantage that a final control element of a motor is relatively bulky. Hydraulic control systems have the advantages that small actuators can produce large forces or torques, components are more rugged and resistant to shock and vibration than electrical ones, but have the disadvantages of requiring a pressurised hydraulic fluid with supply and return lines, the danger of oil leaks, and expense. Pneumatic systems have the advantages of being safe if leaks occur, rugged and easily maintained, but have the disadvantages of requiring a source of compressed air, are slower acting (because air is compressible) and because of the lower pressures involved give a smaller output power than hydraulic systems.

The above can be considered to be just the conceptual design, i.e. just a way of getting some idea of the various elements that would be needed to give the required system. What then needs to follow is a representation of the conceptual design by a mathematical model so that its behaviour can be analysed, e.g. its response to a sudden change, its speed of response and its stability. Only after such a study would the system be constructed and tested. The development of such mathematical models requires mathematics beyond that assumed for this book, so only clues as to their development are given here. For further information the reader is referred to more specialist texts, e.g. *Control Systems Engineering and Design* by S. Thomson (Longman 1989) or *Design of control systems* by A. F. D'Souza (Prentice Hall 1988).

The steps following the conceptual design are thus likely to be:

6 Develop a mathematical model, i.e. an equation or set of equations, which can describe how each of the elements behaves and hence how the entire system will behave.
7 Analyse the behaviour of the mathematical model, considering such factors as response to changes in set values, responses to disturbances, speed of response, and stability.
8 Modify the model to arrive at the required system behaviour.
9 Then construct the actual control system.
10 Test the system to ensure that it meets the required performance criteria.

Example 1

Suggest a pneumatic system that could be used for the control of the temperature of a room. High accuracy is not required with the temperature of the room being generally about 20 °C. A visual indication of the temperature of the liquid is to be given by a pointer–scale arrangement.

Answer

The variable to be controlled is the temperature of a room. Since pneumatic systems are good at opening and closing valves the property to be manipulated could be the flow of hot water through a pipe. This pipe could then be connected to radiators in the room. The heating effect would then be determined by the rate at which the hot water passes through the pipe. The measurement system needs to respond to temperatures of about 20 °C. High accuracy is not required. The output from the measurement system has to be combined with the set point signal to give a pneumatic error signal which can be used via an actuator to operate the control valve. One possibility for the measurement transducer is a bimetallic strip. Movement of this could be used to drive a pointer across a scale and also to move a flapper for a pneumatic-flapper controller. The resulting pressure output could be used to drive an actuator and hence control the opening of the control valve. Figure 9.1 shows an outline of the proposed system. A slightly more ambitious controller would be to use the pneumatic proportional controller described in Fig. 5.13.

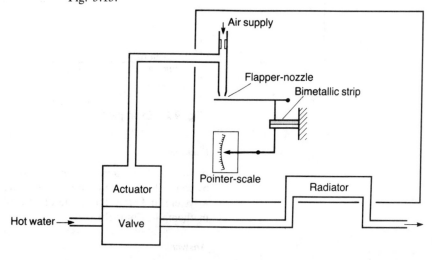

Fig. 9.1 Example 1

Example 2

Suggest an electrical control system that could be used to control the temperature of an oven. The temperature required is 300 °C and the temperature has to be maintained at this value to an accuracy of ± 5 °C.

Answer

The variable to be controlled is the temperature of an oven, the required accuracy being ± 5 °C for a temperature of 300 °C. Since an electrical system is specified, the obvious way of changing the temperature is via an electrical heater. Hence a system is required for the control of the current through the heater. An electrical measurement system which can be used at 300°C and give the required accuracy is a resistance thermometer. This would need a Wheatstone bridge as signal conditioner. The out-of-balance signal from the bridge could be the feedback signal from the measurement system. This could be combined with the set value signal by a differential amplifier. The output from this would need power amplification in order to be large enough to operate the heater. Figure 9.2 shows an outline of the system.

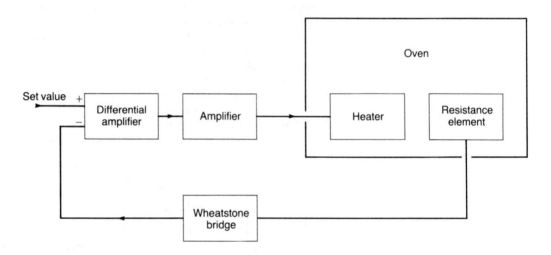

Fig. 9.2 Example 2

Example 3

A simple float system is to be used to control the liquid level in a tank by controlling the amount of liquid entering the tank. (*a*) Devise such a system and state the mode of control adopted, and (*b*) propose a mathematical model for it.

Answer

(*a*) Figure 2.17 shows a possible form for the control system. Movement of the float results in a beam rotating about a pivot and so via the actuator adjusting the opening of the valve. The set point can be adjusted by raising or lowering the connection point of the float to the beam. The error signal is the difference between the position of the float when at the required level and when it is at any other level. The controller is the pivoted beam with an input at the float end of the error signal and an output at the point where the actuator is connected. The output is proportional to the input and so the arrangement is a proportional controller.

$$\frac{\text{Output}}{\text{input}} = \frac{\text{output to pivot distance}}{\text{input to pivot distance}}$$

(b) The pivoted arm is a proportional controller and so has a transfer function K_1 which is independent of time. We might assume that the actuator-valve system has also a transfer function K_2 which is independent of time. The process is likely to have a transfer function of the form $K_3/(\tau s + 1)$ since it is a first-order system. Thus the closed-loop transfer function will be (see Ch. 2 for similar derivations)

$$h_o(s) = \frac{K_1 K_2 K_3/(\tau s + 1)}{1 + K_1 K_2 K_3/(\tau s + 1)} h_i(s)$$

$$+ \frac{K_3/(\tau s + 1)}{1 + K_1 K_2 K_3/(\tau s + 1)} d(s)$$

$$= \frac{K_1 K_2 K_3}{\tau s + 1 + K_1 K_2 K_3} h_i(s)$$

$$+ \frac{K_3}{\tau s + 1 + K_1 K_2 K_3} d(s)$$

We can use this model to determine what the effect of, say, an abrupt change in the set value would be, i.e. a step input.

$$h_o(s) = \frac{K_1 K_2 K_3}{\tau s + 1 + K_1 K_2 K_3} h_i(s)$$

This equation can be simplified if we define two constants a and K as

$$a = \frac{1 + K_1 K_2 K_3}{\tau}$$

$$K = \frac{K_1 K_2 K_3}{1 + K_1 K_2 K_3}$$

Hence

$$h_o(s) = K \frac{a}{s + a} h_i(s)$$

A step input for $h_i(s)$ has the Laplace transform $1/s$, hence

$$h_o(s) = K \frac{a}{s(s + a)}$$

The equation which would give this transform (see the Appendix) is

$$h_o = K(1 - e^{-at})$$

This is an exponential growth to reach a steady state value of K.

Control system performance

In considering the performance required and given by a control system a number of criteria are used. These are:

1 *The steady state accuracy* – how accurately the system responds to a change in input.

2 *The time response* – how the system response to an input change varies with time. This is often referred to as the transient response.

3 *The frequency response* – how the system response to a sinusoidal signal depends on its frequency.

Steady state accuracy

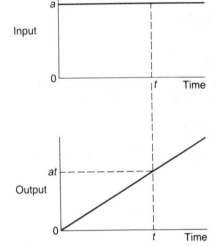

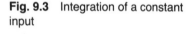

Fig. 9.3 Integration of a constant input

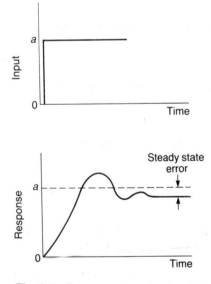

Fig. 9.4 Response to a step input of a type 0 system

The steady state accuracy is the accuracy with which a control system tracks the input set signal. Factors which can affect the steady state accuracy of a control system include non-linearity errors in the measurement system, tolerances on components, and hysteresis, dead-bands, frictional effects, etc. on controller components. The accuracy also depends on how the error signal is processed by the system. Systems can be classified into three groups. A type 0 system has no integrator in the forward-path transfer function. A type 1 system has one integrator in the forward-path transfer function. A type 2 system has two integrators in the forward path transfer function.

An integration exists in the plant process if a constant input to the system results in a ramp change in the variable (Fig. 9.3). This is because the integral of the constant input between zero time and time t, i.e. the area under the input–time graph between 0 and t, is proportional to the time. A ramp form of output is just one where the output is proportional to time.

$$\text{Output} = \int_0^t a\,dt = at$$

where a is a constant. An example of such a plant process is where there is a flow of liquid into a tank in order to control the liquid level. A constant inflow rate results in a level increase which is proportional to time.

In Chapter 5 the different ways were discussed by which control action can be exerted by a controller. Thus a proportional only controller would not include any integral action while a proportional plus integral controller would include one integrator.

The accuracy depends on not only the type of system but also the form the input takes. With a step input of height a a type 0 system will give an error of

$$\text{error} = \frac{a}{1 + k}$$

where k is the forward-path transfer function. Figure 9.4 shows how the response of such a system changes when there is a step input. With a type 1 or 2 system, a step input gives no error.

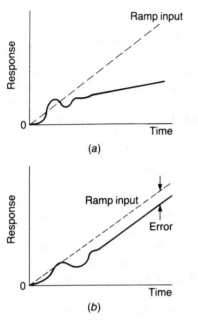

Fig. 9.5 Response to a ramp input of (a) a type 0 system, (b) a type 1 system

With a ramp input (Fig. 9.5) a type 0 system cannot be used, a type 1 gives an error of

$$\text{error} = \frac{\text{ramp gradient}}{k}$$

while a type 2 system gives no error.

Example 4

An electric motor rotates at a constant rate when supplied with a constant current input and is used in a control system to control the angular displacement of a shaft. Does the process include an integrator

Answer

The input is constant and the output, the angular displacement, is proportional to time. The process thus includes an integrator.

Example 5

As part of a control system an electric motor is used to control the angular position of a radar tracker. The controller uses proportional plus integral control. What will be the steady state error if the system is used to track an object which has a steadily increasing angular displacement?

Answer

Because there is integral action in the process, see previous example, and integral action in the controller the system can be classified as type 2. The input signal is a ramp signal and thus there is no steady state error. If the controller had been only proportional the system would have been type 1 and a steady state error would have occurred. Such an error would have meant that the tracker alignment always lagged behind the required angular displacement.

Example 6

What will be the steady state error for the level control system described by Fig. 2.17 if there is a sudden increase in the rate at which water enters the tank to a new constant rate?

Answer

The controller is a simple lever mechanism and so the displacement of the actuator is proportional to the level. The controller is thus just a proportional controller. The process has an input of rate of flow of water and an output of level. This is an integrator. The system is thus a type 1. Hence when there is a step input there is no steady state error. If the water flow had been increasing at a steady rate, i.e. a ramp input, then there would have been a steady state error of ramp gradient/forward path transfer function.

Transient response

The transient behaviour of a system is what happens when there is a change in the set point or a change produced by a disturbance. Thus in the case of a domestic central-heating system the transient behaviour is what happens when the set point of the thermostat is changed from, say, 18 °C to 20 °C, or what happens when a window is opened and a cold blast of air enters. A knowledge of such behaviour is necessary if the quality of the control system is to be judged.

A test procedure that is used is to determine the behaviour of a system when there is a step input. The result depends on the order of the system (see Ch. 1 and Fig. 1.9). With a first-order system the variable changes exponentially with time and gradually reaches the steady value. With a second-order system the result depends on the degree of damping in the system. With damping greater than the critical damping there is, like the first-order system, no overshoot and oscillations. With less than the critical damping there is an overshoot and oscillations.

The performance criteria used to specify the transient performance are (Fig. 9.6):

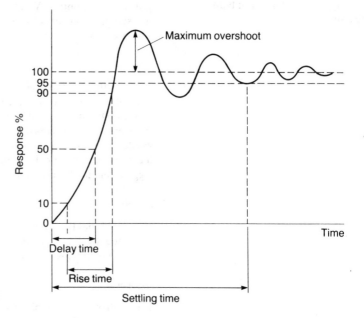

Fig. 9.6 Transient performance terms

1 *Rise time* The rise time is the time taken for the response to a step input to rise from 10 to 90% of its final steady value.
2 *Delay time* The delay time is the time taken for the response to the step input to reach 50% of its final steady value.
3 *Settling time* The settling time is the time taken for the

response to the step input to settle within a certain percentage, usually 5%, of the final steady value.

4   *Maximum overshoot* The maximum overshoot is the maximum deviation, above the final steady value, of the response to the step input. When expressed as a percentage

$$\text{percent max. overshoot} = \frac{\text{max. deviation} - \text{final value}}{\text{final value}} \times 100\%$$

Where it is necessary that a system should not have any overshoot then a first-order system or higher-order system with critical or overdamping is required. Overdamping results in long rise and delay times. Where a quick response is required then some degree of overshoot and oscillation might be suffered and thus an underdamped system used.

Example 7

Suggest the level of damping that should be used with the control system used for the positioning of a robotic arm.

Answer

The robotic arm needs to move, in response to an input signal, to the required position without overshooting and oscillation. If this is to be done in the minimum amount of time then critical damping should be used.

Frequency response

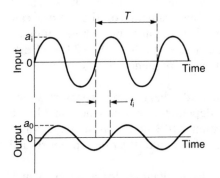

Fig. 9.7   Sinusoidal input and response

The transient response described above is what happens to a system when it is subject to a step input test signal. Another important test input is a sinusoidal signal. When such an input is first applied there are likely to be some transient responses. However when all the transient responses have died down the steady state response is determined. This steady state response is called the frequency response.

Figure 9.7 shows the types of input and output that might be obtained. There can be a change in the amplitude of the sinusoidal signal and a change in phase. The ratio of the amplitudes is called the *frequency response function*.

$$\text{Frequency response function} = \frac{\text{output amplitude}}{\text{input amplitude}}$$

For the example given in Fig. 9.7 the frequency response function is a_i/a_o. The phase difference ϕ in the Fig. is a lag of $(t_i/T) \times 360°$.

The test procedure involves taking measurements of the frequency response function and phase difference for a range

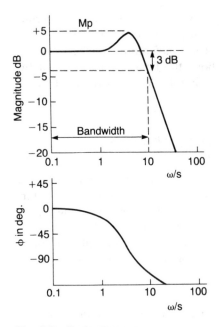

Fig. 9.8 Bode diagram

of different frequencies, starting from the very low. One way of presenting the results is in what is termed a *Bode diagram*. For such a diagram the frequency response is plotted in decibels, i.e. 20 times the log of the frequency response (this is generally referred to as the magnitude), against the log of the angular frequency ω, this being $2\pi f$. Also the phase angle is plotted against the log of the angular frequency. Figure 9.8 shows a typical result.

Two performance criteria that can be derived from a Bode diagram are:

1 *Bandwidth* The bandwidth is the spread between the frequencies for which the frequency response function falls below $-3\,\text{dB}$. In many cases this spread will be upwards from zero frequency.
2 *Peak resonance* The peak resonance M_p is the maximum value of the frequency response function.

A large bandwidth for a control system may result in the system transmitting a lot of unwanted disturbance signals, e.g. electrical noise or mechanical vibrations, round the control loop. Thus a smaller bandwidth which eliminates these signals can be desirable. However a small bandwidth is associated with a system that responds only slowly to changes. Hence a requirement for a fast responding system might well lead to problems with disturbance signals. A large value of peak resonance tends to correspond with a large overshoot with a step input.

The form of the Bode diagram is related to the transfer function of the system. Figure 9.9 shows examples of Bode diagrams for different forms of transfer function. All the magnitude diagrams can be considered to be made up of straight lines, asymptotes. The gradients of these lines can only take certain values. They must be whole number multiples of $\pm 20\,\text{dB/decade}$, i.e. $+40$, $+20$, 0, -20, $-40\,\text{dB/decade}$. A decade is an angular frequency change by a factor of 10, e.g. 1.0 to 10/s.

Figure 9.9(*a*) shows the Bode diagram for a system which has a transfer function independent of frequency and has a magnitude which is a constant 20 dB and a phase angle which is a constant 0°. Such a diagram could describe a proportional controller.

Figure 9.9(*b*) shows the Bode diagram for a system which has a transfer function whose magnitude decreases with increasing frequency at the rate of 20 dB/decade and a phase angle which is constant at 90°. Such a diagram describes a system which is an integrator (see earlier this chapter for discussion of integration). The transfer function is $1/s$ (see Ch. 1 for a discussion of Laplace transforms and the Appendix).

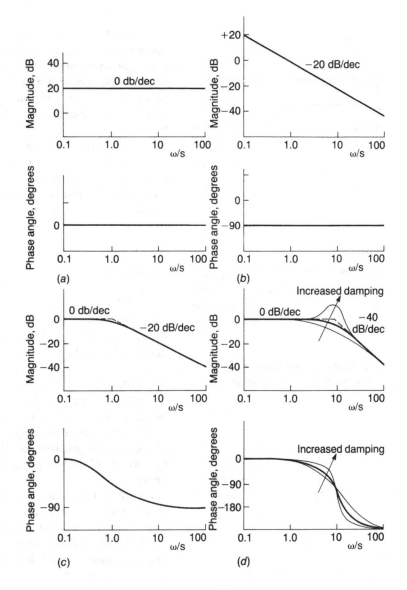

Fig. 9.9 Bode diagrams where the transfer function is (a) constant, independent of frequency, (b) $1/s$, (c) $1/(s\tau + 1)$, (d) $1/(s^2\tau^2 + 2\zeta s\tau + 1)$

Figure 9.9(c) shows the Bode diagram for a first-order system where the transfer function is $1/(s\tau + 1)$. The asymptote has an initial gradient of 0 dB/decade which then changes at the angular frequency 1.0/s to -20 dB/decade. The angular frequency at which the change is considered to take place is known as the *corner frequency* ω_c and is where the asymptotes can be considered to meet. The time constant τ is $1/\omega_c$. Thus the higher the corner frequency the smaller the time constant and hence the quicker the system reacts. However, the higher the corner frequency the greater the bandwidth. Hence a large bandwidth is associated with a fast-reacting system.

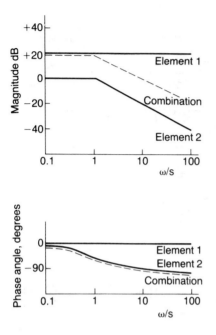

Fig. 9.10 Bode diagram for a two-element system, proportional plus first order

Figure 9.9(*d*) shows the Bode diagram for a second-order system with a transfer function of $1/(s^2\tau^2 + 2\zeta s\tau + 1)$, where ζ is the damping ratio (it has a value of 1 at critical damping). Such a system has an initial asymptote of gradient 0 dB/decade which at the corner frequency changes to -40 dB/decade. As before, the time constant τ is the reciprocal of the corner frequency.

The Bode diagram can be used to describe the behaviour of a single element in a control or measurement system or the entire system. Suppose we have a system consisting of just two elements. The transfer function of the combination will be the product of the transfer functions of the two separate systems (see Ch. 1).

$$G = G_1 G_2$$

Hence

$$\log_{10} G = \log_{10} G_1 + \log_{10} G_2$$

Thus the magnitude on a Bode diagram of the combined transfer function can be obtained by adding together the magnitudes of the two separate transfer functions. Figure 9.10 shows this for a combination of a proportional element with a first-order element having a transfer function of $1/(s\tau + 1)$. The phase angle for the combination is just the sum of the phase angles for the separate systems.

The effect of combining a proportional element with other elements is to just increase the overall transfer function of the system at all frequencies. There is no effect on the phase angle.

Figure 9.11 shows the result of combining an integral element with a first-order system having a transfer function $1/(s\tau + 1)$. The combined system has now an initial gradient of 20 dB/decade. The low-frequency magnitude has been increased and the phase angles all reduced.

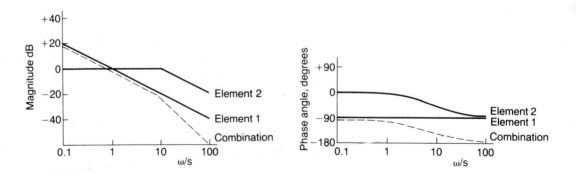

Fig. 9.11 Bode diagram for a two-element system, integral plus first order

A system with no integration element is called a type 0 system (see earlier this chapter). The Bode diagram for such a system will thus have a zero gradient at low frequencies. A type 1 system with one integration will have a low frequency slope of -20 dB/decade, as in Fig. 9.11). A type 2 system with two integrations will have a low-frequency slope of -40 dB/decade.

For a more detailed discussion of the Bode diagram and how transfer functions can be derived from it, or Bode diagrams derived from transfer functions, the reader is referred to more specialist texts, e.g. *Design of control systems* by A. F. D'Souza (Prentice Hall 1988).

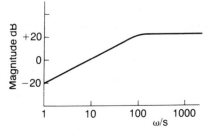
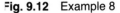

Fig. 9.12   Example 8

Example 8

Estimate the bandwidth for the system giving the Bode diagram shown in Fig. 9.12.

Answer

The bandwidth is from an angular frequency of about 100/s upwards, i.e. a frequency of $100/2\pi = 16$ Hz upwards. Such a system filters out the low frequencies and is responding only to high frequencies.

Example 9

What is the time constant for the system giving the Bode diagram shown in Fig. 9.9(*c*)?

Answer

The corner angular frequency is about 1.0/s. Hence the time constant is the reciprocal of this frequency and so is 1 s.

Example 10

For the Bode diagram shown in Fig. 9.9(*a*) what is the value of the transfer function?

Answer

The Bode diagram has a zero gradient and so a constant value of 20 dB. This means

$$20 \log_{10}G = 20$$

Hence the transfer function G has the value 10.

Example 11

The Bode diagram of a system is found to have a magnitude which has a low-frequency gradient of -20 dB/decade. Will there be steady state errors with such a system when it is subject to (*a*) a step input, (*b*) a ramp input?

Answer

Such a system will contain one integration and so is a type 1 system. With such a system there will be no steady state error with the step input but one with the ramp input.

Fault finding

With a measurement system or a control system there are essentially three steps that should be followed in establishing the location of a fault. They might be stated as understanding, reasoning and finally testing.

1   Establish how the system operates. This means finding out what its constituent elements are and their functions. Find any manuals or specifications for the system and its elements.
2   Pin-point the fault. Try without using any test equipment to establish from the nature of the fault where in the system it can be located.
3   The final step is to use test equipment.

Problems

1   A company is considering developing new, modern, control systems for use with motor vehicles. You have been asked to present a report outlining the current methods used and presenting suggestions for other possibilities. Consider the systems used to control (*a*) the engine temperature, (*b*) the fuel intake to the cylinders of the engine.
2   Outline the possibilities for a control system that could be used to sound an alarm bell if a lift is overloaded.
3   Outline the possibilities for a control system that could be used to control the level of water in a tank. The system is required to maintain the level by adjusting a valve which controls the amount of water leaving the tank.
4   Suggest the elements that could be used for the following control systems:
    (*a*) control of the rate of flow of liquid along a pipe,
    (*b*) control of the pressure in a pressurized tank,
    (*c*) control of the weight of sand poured into a lorry via a chute,
    (*d*) a machine tool to cut material in the shape dictated by a template,
    (*e*) rotation of a small wheel by hand to be copied in the rotation of a much larger wheel.
5   An electro-hydraulic control system is used in a machine tool to control the position of the tool. The input commands are generated by a probe following a template. If the system can be classified as type 1, what will be the steady state error when it is subject to (*a*) a step input, (*b*) a constantly increasing input?
6   Sketch the Bode diagram for a system having a transfer function of $1/(s\tau + 1)$ and a time constant of $10\,\text{s}$.
7   What will be the form of the Bode diagram of a system having a bandwidth of $0\text{–}500\,\text{Hz}$?

8 What will be the form of the Bode diagram of a system having a transfer function of 100 which is independent of frequency?

9 A system has a steady state error when subject to a step input. How can such an error be eliminated and what will be the effect on the Bode diagram for the system?

Appendix

Laplace transforms

Laplace transforms enable differential equations to be transformed into equations which can then be handled as simple algebraic equations. The following are some of the basic operations involved in transforming a differential equation and which would be adequate to deal with the equations met in this book. It has been assumed that the function concerned has the value zero for all times before $t = 0$.

1   The addition of two functions becomes the addition of their two Laplace transforms.

$$f_1(t) + f_2(t) \text{ becomes } F_1(s) + F_2(s)$$

2   The subtraction of two functions becomes the subtraction of their two Laplace transforms.

$$f_1(t) - f_2(t) \text{ becomes } F_1(s) - F_2(s)$$

3   The multiplication of some function by a constant becomes the multiplication of the Laplace transform of the function by the same constant.

$$af(t) \text{ becomes } aF(s)$$

4   The first derivative of some function becomes s times the Laplace transform of the function.

$$\frac{\mathrm{d}}{\mathrm{d}t}f(t) \text{ becomes } sF(s)$$

5   The second derivative of some function becomes s^2 times the Laplace transform of the function.

$$\frac{\mathrm{d}^2}{\mathrm{d}t_2}f(t) \text{ becomes } s^2F(s)$$

6   The nth derivative of some function becomes s^n times the Laplace transform of the function.

$$\frac{\mathrm{d}^n}{\mathrm{d}t_n} f(t) \text{ becomes } s^n F(s)$$

7   The first integral of some function, between zero time and time t, becomes $(1/s)$ times the Laplace transform of the function.

$$\int_0^t f(t) \text{ becomes } \frac{1}{s} F(s)$$

The following are some of the more common Laplace transforms and their corresponding time functions.

Laplace transform	Time function	
1		A unit impulse
$\dfrac{1}{s}$		A unit step function
$\dfrac{1}{s^2}$	t	A ramp function of unit slope
$\dfrac{1}{s^3}$	$\dfrac{t^2}{2}$	
$\dfrac{1}{s + a}$	e^{-at}	Exponential decay
$\dfrac{1}{(s + a)^2}$	te^{-at}	
$\dfrac{a}{s(s + a)}$	$1 - e^{-at}$	Exponential growth
$\dfrac{a}{s^2(s + a)}$	$t - \dfrac{(1 - e^{-at})}{a}$	
$\dfrac{s}{(s + a)^2}$	$(1 - at)e^{-at}$	
$\dfrac{\omega}{s^2 + \omega^2}$	$\sin \omega t$	Sine wave
$\dfrac{s}{s^2 + \omega^2}$	$\cos \omega t$	Cosine wave
$\dfrac{\omega}{(s + a)^2 + \omega^2}$	$e^{-at} \sin \omega t$	
$\dfrac{s + a}{(s + a)^2 + \omega^2}$	$e^{-at} \cos \omega t$	

For more details of Laplace transforms the reader is referred to more specialist texts, e.g. *Control Systems Engineering and Design* by S. Thompson (Longman 1989) or *Design of Control Systems* by A. F. D'Souza (Prentice Hall 1988).

Answers to problems

Chapter 1

2. (a) $\pm 1\,\text{mA}$, (b) $\pm 2\,\text{V}$
3. $4.0 \pm 0.1\,\text{litre/s}$
4. $40\,\text{mm}$
5. $0.83\,\text{mV/kPa}$
6. $5\tau = 25\,\text{s}$
7. (a) No change, (b) exponentially reaches 12 after about 20 s, (c) underdamped oscillation, $\zeta = 0.5$
8. $0.40\,\text{mm/Pa}$
9. $9.6\,\text{mm/°C}$
10. $\pm 4\%$

Chapter 2

7. $0.14\,\text{mA/°C}$
8. (a) $2.0\,\text{mA}$ per m/s, (b) $0.6\,\text{m/s}$ per mA, (c) $0.3\,\text{mA}$ per m/s
9. (a) $K_1 K_2/(1 + K_1 K_2 K_3)$, (b) $K_1 K_2/(\tau s + 1 + K_1 K_2 K_3)$, (c) $K_1 K_2/(\tau s^2 + s + K_1 K_2 K_3)$

Chapter 3

1. (a) temperature–e.m.f, (b) light intensity–resistance change, (c) strain–resistance change, (d) displacement–mutual inductance change–voltage, (e) flow rate–pressure difference.
2. For example, (a) LVDT, (b) spring, (c) load cell with strain gauges, (d) optical digital shaft encoder, (e) resistance thermometer
3. $-7.9\ \%$
4. $+1.54\,°\text{C}$
6. $0.21\,\Omega$
7. 0.73%
8. $11.7\,\text{mm}$

Chapter 4

1. $8.0\,\Omega$
2. $0.066\,V$
3. $1.22\,\mu A$
5. $0.059\,V$
6. $5.25 \times 10^{-5}\,V$
8. $2.4\,\mu F$, $0.83\,k\Omega$
9. $2.49 \times 10^{-8}\,F$
10. 33
11. $R_2/R_1 = 20$
13. 9
15. Possibly (a) load cell, strain gauges, Wheatstone bridge, differential amplifer, sample and hold, analogue to digital, counter. (b) thermistor, bridge or simple series circuit, voltage-to-frequency converter, multiplexer, sample and hold, analogue to digital, counter.

Chapter 5

3. (a) 8 minutes, (b) 20 minutes
4. (a) 12 s, (b) 24 s
6. 20%, 5%/%
7. 62.5%, 0.63%
8. (a) 51.0%, (b) 51.0%, (c) 0, (d) 49.5%
9. (a) 54%, (b) 66%
10. (a) 60%, (b) 73%, (c) 76.1%, (d) 62%
13. 3%/%, 0.0025%/% s, 100% s/%
14. 3%/%, 0.01%/% s, 25% s/%

Chapter 6

5. (a) $0.05\,m^3/s$, (b) $0.10\,m^3/s$
6. (a) $0.42\,m^3/s$, (b) $0.89\,m^3/s$
7. $960\,mm$

Chapter 7

5. 0.5, $\times 0.73$
6. $60\,\Omega$ in series

Chapter 9

5. (a) No error, (b) ramp gradient/forward transfer function
6. Magnitude zero gradient to 0.1/s, then $20\,dB$/decade
7. Zero gradient up to ω of about 3000/s
8. Zero gradient at magnitude $40\,dB$
9. Add integration, low-frequency gradient $-20\,dB$/decade

Index

a.c bridges, 75
Accuracy, 3
Accuracy of combined system
    elements, 16
Acquisition time, 92
Active device, 42
Actuators, 121, 122
Alarm indicators, 137
Amplification, 1, 79
Amplifier, electronic, 82
Amplifier, gear train, 80
Amplifier, lever, 80
Amplifier, operational, 82
Amplitude modulation, 88
Analogue chart recorders, 138
Analogue display, 132
Analogue instruments, 19
Analogue to digital conversion, 89
Aperture time, 92
Attenuation, 86

Bandwidth, 6
Bellows, 14, 60, 169
Bernoulli's equation, 62
Bimetallic strip, 97, 159
Bit, 19
Bode diagram, 190
Bourdon tube, 61, 168
Bubbler method, 174

Calibration, 20
Capacitance transducers, 45, 62,
    175
Capsules, 60, 169
Cascade control, 114
Cathode ray oscilloscope, 146
Cathode ray tube, 146
Chart recorders, see recorders
Closed-loop control systems, 25, 26
Closed-loop differential pressure
    cell, 169
Closed-loop recorder, 145
Closed-loop transfer function, 30
Comparison element, 26, 95
Continuity equation
Control element, 27
Controllers, 95–120
Control systems, 24–41, 181–95
    closed-loop, 25, 26
    designing, 181
    error, 186
    frequency response, 189
    open-loop, 25
    performance, 185
    steady state accuracy, 185
    transient response, 188
Control valve, 126
Control valve sizing, 130

Conversion time, 90
Correction element, 27, 121
Critical damping, 10, 141
Cross-talk, 186
Current standard, 21
Current to pressure converter, 85

Damping factor, 141
Dall flowmeter, 63, 172
Data display, 132–58
Data logger, 156
Dead space, 4, 147
Decibel, 7, 19
Derivative control, 105
De Souty bridge, 77
Diaphragms, 60, 169
Diaphragm actuator, 124
Dial indicator gauge, 81
Differential amplifier, 83, 96
Differential gear, 96
Differential pressure
    transducers, 45, 62, 172
Digital control, 116
Digital display, 132
Digital instruments, 19
Digital meter, 134
Digital printers, 185
Digital shaft encoder, 65
Digital to analogue conversion, 91
Dipstick, 174
Displacement measurement, 54, 55
Display element, 1, 132
Disturbances, effect of, 35
Double beam oscilloscope, 148
Dual slope voltmeter, 136
Dumb instruments, 19
Dynamic characteristics, 8, 37

Elastic transducers, 45, 59
Electronic controller,
    derivative, 105
    integral, 109
    proportional, 103
    three mode, 112
Encoders, 65
Equal percentage plug, 129
Error, control systems, 186
Error, defined, 3, 4
Error, sources of, 18
Error, non-linearity, 4

Fault finding, 194
Feedback, 29
    negative, 29
    positive, 29
Filters, 87
First order instrument, 8, 12
Flapper-nozzle transducer, 61

Float systems, 174, 177
Flow measurement, 62, 172
Force measurement, 59, 60
Forward path transfer function, 31
Frequency modulation, 88

Galvanometric recorder, 138
Gas in metal thermometer, 161
Gauge factor, 48
Gear train as amplifier, 80

Holding time, 92
Huggenberger extensometer, 80
Hydraulic amplifier, 125
Hydraulic cylinder, 125
Hydraulic servomotor, 125
Hysteresis, 5

Incremental shaft encoder, 65
Inductance transducers, 44, 54
Integral control, 108
Intelligent instruments, 19
Intermediate temperatures, law
    of, 56
Inverting amplifier, 82
Inverting summer, 96

Knife-edge recorder, 138

Lag, 5, 97
Laplace transform, 11, 196
Law of intermediate
    temperatures, 56
Length standard, 21
Level control, 27, 37, 114
Level measurement, 53, 174
Lever as amplifier, 80
Linear actuator, 125
Linear contoured plug, 128
Linearization, 84
Linear variable differential
    transformer, 55
Liquid in glass thermometer, 160
Liquid in metal thermometer, 160
Load cell, 43, 49, 60, 176
Luminous intensity standard, 21
LVDT, see linear variable differential
    transformer

Magnetic tape recorder, 151
Manometers, 14, 167
Mass, spring, damper system, 10, 14
Mass standard, 20
Maxwell bridge, 76
Measurement systems, 1–21,
    159–78
Measurement system design, 159
Modulation, 87

Monitors, 150
Motors as actuators, 122
Moving-coil meter, 132
Multiplexers, 93
Multistep control, 98

Noise, 19
Non-linearity error, 4
Nozzle flowmeter, 63, 172

Offset, 102
Open-loop control systems, 25
Open-loop transfer function, 32
Operational amplifier, 82
Orifice flowmeter, 63, 172
Overshoot, 10
Owen bridge, 77

Parallax error, 18
Passive device, 42
Phase modulation, 88
Phosphors, 147
Photoconductive cell, 52
Photovoltaic transducers, 44, 59
Piezo-electric transducers, 44, 59, 170
Pitot static tube, 63, 172
Pneumatic controller, proportional, 102
Pneumatic controller, proportional plus derivative, 107
Pneumatic controller, proportional plus integral, 111
Pneumatic controller, three-mode, 112
Pneumatic control valve, 126
Pneumatic cylinder, 125
Pneumatic transducers, 45, 61
Positive displacement meter, 173
Potentiometer, 8, 50, 96
Potentiometer measurement system, 78
Potentiometric recorder, 146
Precision, 4
Pressure, absolute, 167
Pressure, gauge, 167
Pressure measurement, 50, 52, 60, 61, 167
Process control, 24
Process element, 27
Process reaction curve, 117
Proportional band, 100
Proportional control, 100
Proportional plus derivative control, 106
Proportional plus integral control, 110
Proportional plus integral plus derivative control, 112
Proving ring, 59
Pulse amplitude modulation, 87
Pulse duration modulation, 87
Pulse width modulation, 87
Pyrometer, 162

Quick-opening plug, 127

Radiation pyrometer, 162
Ramp type voltmeter, 135
Random errors, 18
Range, 4
Raster display, 150
Recorders, 138
  analogue, 138
  closed-loop, 145
  dynamic behaviour, 139
  galvanometric, 138
  knife-edge, 138
  magnetic tape, 151
  potentiometric, 145
  ultraviolet, 139
Relays as actuators, 122
Reliability, 4
Repeatability, 4
Reproducibility, 4
Resistance, temperature coefficient of, 45
Resistance thermometer, 1, 2, 45, 161
Resistance thermometer compensation, 74
Resistance transducers, 43, 45
Resolution, 4
Response time, 5
Rotameter, 63, 173
Rotating disc transducers, 45, 65
Rotating lobe meter, 173

Sample and hold, 92
Second order instrument, 10, 14
Sensing element, see transducer
Sensitivity, 4
Servo system, 24
Shaft encoder, 65
Signal conditioner, 1, 69
Signal converter, see signal conditioner
Signal to noise ratio, 19
Signal processor, see signal conditioner
Slew rate, 92
Solenoids as actuators, 122
Speed control, 28
Spring balance, 43
Spool valve, 125
Stability, 4
Standards, primary, 20
Standards, secondary, 21
Static characteristics, 8
Stepping motor, 123
Storage oscilloscope, 149
Strain gauge, 48, 83
Strain gauge compensation, 49, 74
Successive approximations voltmeter, 135
Systematic errors, 18

Temperature measurement, 159
Temperature standard, 21

Thermistor, 47
Thermocouple, 42, 45, 62, 162
Thermocouple compensation, 73
Thermoelectric transducers
Thermometer,
  bimetallic, 159
  dynamic characteristic, 9, 13
  gas in metal, 161
  liquid in glass, 160
  liquid in metal, 160
  radiation, 162
  resistance, 1, 2, 42, 161
  thermoelectric, 42, 45, 62, 162
  vapour pressure, 161
Thermopile, 58
Thermostat, 97
Three mode controller, 112
Threshold, 5
Time constant, 12
Time standard, 21
Transducer, defined, 1
Transcucer(s). 42–68
  capacitance, 44, 52
  differential pressure, 45, 62
  elastic, 45, 59
  inductance, 44, 54
  photovoltaic, 44, 59
  piezo-electric, 44, 59
  pneumatic, 45, 61
  resistance, 43, 45
  rotating disc, 45, 65
  thermoelectric, 44, 56
  turbine, 45, 64
Transducers, types of, 42
Transfer function,
  closed-loop, 30
  defined, 6
  effect of disturbances, 35
  forward path, 31
  multi-loop, 33
  of series elements, 15
  open-loop, 32
  system, 15
Tuning, 117
Turbine transducers, 45, 62
Two-colour pyrometer, 164
Two-step control, 97

Ultimate cycle method, 118
Ultrasonic level gauge, 176
Ultraviolet recorder, 139

Valve, control, 126
Vapour pressure thermometer, 161
Variable area flowmeter, 63
Variable differential inductor, 54
Variable reluctance transducer, 54
Venturi tube, 62, 172
Voltage to current converter, 85
Voltage to frequency converter, 89
Voltage to frequency voltmeter, 135

Wheatstone bridge, 69

Wheatstone bridge, deflection
    type, 70
Wheatstone bridge, strain gauge
    compensation, 74
Wheatstone bridge, thermocouple

compensation, 73
Wheatstone bridge, thermometer
    compensation, 74
Wien bridge, 77
Word, 19

Zero drift, 5
Zero order instrument, 8
Ziegler and Nichols, 117

LIVERPOOL
JOHN MOORES UNIVERSITY
AVRIL ROBARTS LRC
TITHEBARN STREET
LIVERPOOL L2 2ER
TEL. 0151 231 4022